Ring-disc

Electrodes

W. J. Albery

and

M. L. Hitchman

CLARENDON PRESS OXFORD 1971

Oxford University Press, Ely House, London W.1

GLASGOW NEW YORK TORONTO MELBOURNE WELLINGTON
CAPE TOWN SALISBURY IBADAN NAIROBI DAR ES SALAAM LUSAKA
ADDIS ABABA BOMBAY CALCUTTA MADRAS KARACHI LAHORE DACCA
KUALA LUMPUR SINGAPORE HONG KONG TOKYO

Set by E.W.C. Wilkins & Assoc. Ltd.,
and printed in Great Britain by
Lowe and Brydone Ltd., London

To   R.E.R.

PREFACE

WHEN we checked our manuscript we found that nearly every chapter either started or ended with the words 'the ring-disc electrode is a useful technique .....'. This is the reason why we have written this book and we make no apologies for it. We have described the applications of the electrode, not only to the elucidation of electrode mechanisms, but also to the study of homogeneous reactions, to analytical measurments, and to the understanding of transient behaviour. We have given the outlines of the mathematical theory because the great advantage of the ring–disc electrode is that in many cases one can calculate exact solutions for the flow of material from one electrode to the other. Our work at Oxford has been concerned with developing this theory and testing it by experiment. We feel that the theory is now becoming well established and we look forward to the application of the ring–disc electrode to many problems in electrochemistry, reaction kinetics, and analysis.

There are many people we would like to thank but first and foremost are our fellow research workers and collaborators at Oxford, Jens Ulstrup, Mary Archer, Nigel Field, Roger Cook, Jon Drury, and Michael Wormald. Then we would like to thank Professor R.P. Bell under whose cigar smoke the first electrodes were rotated in Oxford. Next, we thank the Warden of Merton, not as Warden of Merton, but as Rex Richards, for it was in his time as Dr. Lee's Professor that this work was carried out. We have had many interesting and stimulating conversations with other workers in the field, but our conversations with Professor Bruckenstein whether at Bernie's in Minneapolis or in a punt on the Cherwell deserve special mention. We also thank the organizers of the Gordon Research Conference on Electrochemistry for the exciting and valuable meeting at Santa Barbara in January 1969.

Judging by her decipherment of our manuscript Mrs. Elizabeth Price could do good work on the Dead Sea scrolls and we commiserate with Mrs. Pauline Hitchman for becoming a ring-disc widow. Finally we thank University College Oxford, Wolfson College, and I.C.I. for research fellowships that we have held during the course of this work, and the Science Research Council for a grant for equipment.

W.J.A.
M.L.H.

CONTENTS

LIST OF SYMBOLS

a	Concentration of A
a	Geometrical parameter, eqn (9.10)
A	Area of electrode
A_1	$3\frac{1}{3}\Gamma(4/3) = 1\cdot288$
A_2	$0\cdot643\,\nu^{1/6}D^{1/3}$
$Ai(\ )$	Airy function
b	Concentration of B
b	Geometrical parameter, eqn (9.10)
B,B'	Geometrical parameters in eqn (10.8)
c	Concentration of C
c	Geometrical parameter, eqn (9.10)
C	$0\cdot510\,\omega^{3/2}\,\nu^{-1/2}$, convection constant
D	Diffusion coefficient
e	Electron
E	Potential
E_D	Potential of the disc electrode
E_R	Potential of the ring electrode
f	Dimensionless flux function on the disc, eqn (10.3)
fr	Frequency in Hz
F	Faraday
$F(\gamma)$	Function describing radial flow, Fig. 2.3
$F(\ )$	Various functions
$\mathcal{F}(\ )$	Hypergeometric function
g	Dimensionless concentration gradient at electrode surface
g	Dimensionless concentration function for second-order kinetics

$G(\gamma)$ Function describing axial flow, Fig. 2.3

h Dimensionless concentration function for second-order kinetics

$H(\gamma)$ Function describing flow normal to disc electrode, Fig. 2.3

"i" $\sqrt{-1}$

i Current

i_D Disc current

$i_{D,L}$ Transport limited current at disc electrode

i_L Transport limited current

i_R Ring current

$\bar{i}_R$ Laplace transform of i_R with respect to t

$i_{R,L}$ Limiting ring current

$i_{R,L}^0$ Limiting ring current when $i_D = 0$

j Flux (mol cm^{-2} s^{-1})

j_L Transport limited flux

J $4\rho/\pi$ for reversible system

k Homogeneous rate constant

k' Heterogeneous rate constant (cm s^{-1})

k'_* Particular heterogeneous rate constant (cm s^{-1}) for one square when $k'_{-1} = k'_2$

k'' First-order heterogeneous rate constant for H$^+$ transfer (cm s^{-1})

k''' Second-order heterogeneous rate constant for H$^+$ transfer (cm^4 mol^{-1} s^{-1})

K Equilibrium constant

L Avogadro's number

$\mathcal{L}$ Laplace transformation

M Critical disc current in titration curve, eqn (7.5)

M' $M/(c_\infty nF)$

n Number of electrons in a reaction

N_0 Simple steady-state collection efficiency

N' Collection efficiency at any point on a titration curve

N_0'' Approximate value for N_0 obtained from expression for thin-gap thin-ring electrodes

N_K Collection efficiency for system with first-order kinetics

N_K'	Collection efficiency for system with second-order kinetics when reaction surface is on inside edge of ring electrode
N_K''	Approximate value of N_K obtained from expression for thin-gap, thin-ring electrodes
N_t	Collection efficiency for any point on a transient, eqn (10.4)
N_σ	N_K with κ replaced by $\sigma^{1/2}$
N_G	Laplace transformation of N_t with respect to real time for a galvanostatic transient
$N_{a.c.}$	Alternating current collection efficiency
P	Dimensionless geometric parameter for second-order kinetics, eqn (8.4)
P	$EF/4\cdot606\ RT$, dimensionless potential variable
P'	$\eta F/4\cdot606\ RT$
Q'	Dimensionless parameter for second-order kinetics, eqn (8.5)
r	Distance in radial direction
r_1	Radius of disc electrode
r_2	Internal radius of ring electrode
r_3	External radius of ring electrode
r_J	Radial distance of reaction zone in titration curve
R	Gas constant
R_{CT}	Charge transfer resistance
R_E	Electrode resistance
R_Ω	Ohmic resistance of solution
s	Laplace transform variables of ξ
t	Time
T	Temperature
u	Dimensionless concentration variable
u_*	Value of u on disc electrode surface
u'	Concentration variable used in first order kinetics
v	Velocity of flow of solution
V	Volume of solution
w	$zC^{1/3}D^{-1/3}$, dimensionless distance variable normal to electrode
W	Rotation speed in Hz

x	Branching factor in electrode mechanisms
X	In phase component in $N_{a.c.}$
y	$(r/r_1)w$, dimensionless distance variable
Y	Out of phase component in $N_{a.c.}$
z	Distance normal to the electrode
Z_D	Thickness of diffusion layer, eqn (2.15)
Z_H	Thickness of hydrodynamic layer, eqn (2.1)
N.H.E.	Normal hydrogen electrode
N.C.E.	Normal calomel electrode
S.C.E.	Saturated calomel electrode
α	Transfer coefficient
α	$(r_2/r_1)^3 - 1$, geometrical parameter for gap
α'	$3\ln(r_2/r_1)$, geometrical parameter for gap
β	$(r_3/r_1)^3 - (r_2/r_1)^3$, geometrical parameter for ring
β'	$3\ln(r_3/r_2)$, geometrical parameter for ring
β_J	$(r_J/r_1)^3 - (r_2/r_1)^3$, describes position of reaction zone on the ring electrode
δ	Phase shift in radians
γ	z/Z_H
$\Gamma(\ )$	Gamma function
η	Overpotential
θ	$K, a - b$, perturbation function for equilibrium
ι	Current density (A cm^{-2})
κ_∞	Specific conductivity
κ	$(k_1/D)^{1/2}(D/C)^{1/3}$, dimensionless kinetic parameter for homogeneous reaction
λ	$(k'/D)(D/C)^{1/3}$, dimensionless kinetic parameter for heterogeneous reaction
λ_1	$\{(k_1' + k_{-1}')/D\}(D/C)^{1/3}$, dimensionless kinetic parameter for heterogeneous reaction
μ	$(D/k_1)^{1/2}$, thickness of reaction layer
ν	Kinematic viscosity
ξ	Dimensionless radial variable
ξ	$\frac{1}{3}\{(r/r_1)^3 - 1\}$

ξ_1'	$= a/3 = \xi_1$ for $r = r_2$
ξ_2	$\frac{1}{3}\{(r/r_1)^3 - (r_2/r_1)^3\}$
ξ_2'	$= \beta/3 = \xi_2$ for $r = r_3$
ξ_3	$\ln(r/r_2)$
$\xi_{1,\kappa}$	$\kappa^3 \ln(r/r_1)$
$\xi_{1,\kappa}'$	$\xi_{1,\kappa}$ with $r = r_2$
$\xi_{2,\kappa}$	$\kappa^3 \ln(r/r_2)$
$\xi_{2,\kappa}'$	$\xi_{2,\kappa}$ with $r = r_3$
ξ_1''	$\ln(r_2/r_1)$
ξ_2''	$\ln(r_3/r_2)$
ρ	$\bar{R}_\Omega/\bar{R}_E$, quantity describing non-uniform current distribution
σ	Laplace transform variable of τ
τ	$tC^{2/3} D^{1/3}$, dimensionless time variable
ϕ	Angle in cylindrical polar coordinates
ϕ	$K_{-2}c - b$, perturbation function for equilibrium
χ	General variable
ψ	Dimensionless concentration variable, see eqn (10.1)
ψ_*	Value of ψ on the disc electrode
ψ'	Dimensionless concentration variable used in transient theory
ω	Rotation speed in rad s^{-1}
ω'	$2\pi fr\, C^{-2/3} D^{-1/3}$, dimensionless frequency

INTRODUCTION

The Standard Random Electrode

'THE currents measured will therefore be distinctly fluctuating and by
no means exactly reproducible.' This quotation is taken from the
instructions for an undergraduate experiment (c; 1922–62) in the
Physical Chemistry Laboratory at Oxford. The experiment was called
'e.m.f.'s of polarization' which, together with the Langmuir trough,
served as condign punishment for recalcitrant undergraduates. It
consisted of placing two copper electrodes in an unstirred solution
of $CuSO_4$ and trying to measure the current at an arbitrary time after
applying various potentials. Only the more deceitful undergraduates
ever completed the experiment; the others left Oxford convinced that
nothing good can ever come from electrochemical measurements.

We can express the problem of this experiment in more quantitative terms:

$$i = f(W, N_L, \partial(JD)/\partial t), \qquad (1.1)$$

where i is the current, W is the number of windows open in the laboratory, N_L is the number of lorries passing in South Parks Road, and
$\partial(JD)/\partial t$ is a complicated periodic function describing the supervision
of a typical junior demonstrator. The reason for the fluctuating currents
and the complicated form of eqn (1.1) is that the current at the copper
electrode is determined by the concentration of cupric ions at the
electrode surface, and this concentration is in turn determined by the
transport of the ions from the bulk of the solution to the electrode. In
general, when a current is passing at an electrode there are these
local variations in the concentration of the reactants and the products
of the electrochemical reaction. The solution close to the electrode is
largely stationary with respect to the electrode, and the reactants and
products have to diffuse across this stagnant layer. Since diffusion is
an inefficient transport process the main concentration changes take
place within the diffusion layer. Outside the layer the concentrations

approach and equal their values in the bulk of the solution. Fig. (1.1) shows a typical pattern of concentrations against distance for a

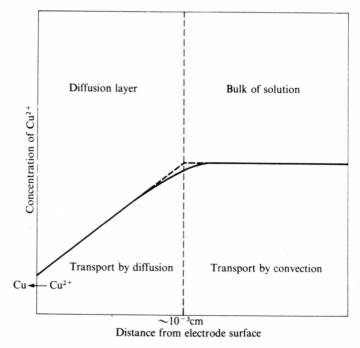

FIG. 1.1. Schematic concentration profile close to the surface of an R.D.E.

rotating disc electrode (R.D.E.) where the following electrode reaction is taking place:

$$Cu^{2+} \xrightarrow{2e} Cu .$$

However, in an unstirred solution the thickness of the diffusion layer is not well defined and the transport depends on the local eddies, vibrations, and swirling motions described in eqn (1.1). To prevent these random fluctuations it is necessary that the stagnant diffusion layer close to the electrode should be smaller than 10^{-2} cm. If the layer becomes larger than this then the chance convective motions in the liquid will become responsible for the transport of species to and from the electrode.

Common Types of Electrode

The current at an electrode is determined by one or more of the following factors:

(1) transport of species to and from the electrode;

(2) kinetics of the electrode process;

(3) kinetics of homogeneous chemical reactions.

In order to obtain reproducible currents it is necessary that the transport of the electroactive species from the bulk of the solution to the electrode and vice versa should be well defined and controlled. But this is not all. In order to separate the influence of the different factors we must not only have a defined system of transport but we must also be able to describe the transport theoretically. Some of the more common types of electrode system are listed in Table 1.1, which also lists the characteristics of each system with respect to time.

TABLE 1.1

Common types of electrode system

Electrode	Well defined transport	Calculable transport	Time characteristics
Stationary	Short times only	Short times only	Transient
Rotating wire	Yes	No	Steady state
Vibrating wire	Yes	No	Cyclic
Mercury drop	Yes	Yes	Cyclic
Rotating disc (R.D.E.)	Yes	Yes	Steady state

The stationary electrode only has defined characteristics for a short time (10^{-1} to 10^2 s) after shifting the potential from the zero current point. At longer times the thickness of the diffusion layer becomes too large. No true steady state can be set up [eqn (1.1)], and in all measurements the current, the electrode potential, or both are varying with time. Rotating and vibrating wire electrodes do have well defined diffusion layers, but we cannot easily calculate the thickness of these from first principles, and therefore these electrodes have to be used in a semi-empirical fashion. Adams[1] gives a useful discussion of these types of electrode. The mercury-drop electrode has been the most widely used system. During the lifetime of each drop it is similar to a stationary electrode with a diffusion layer spreading out from the electrode surface into the bulk of the solution. Before this becomes too large and eqn (1.1) starts to apply, the drop falls off. The current and/or potential varies with time during the life of each drop and the system is therefore a cyclic one. In the rotating disc electrode, as described in Chapter 2, a known pattern of hydrodynamic flow is imposed on the solution giving well defined and calculable transport. Several seconds after any change at the electrode a steady state is established and currents and potential that do not vary with time may be observed. This electrode system is therefore the simplest of those listed in Table 1.1.

Current-Voltage Curve for a Redox System

If there are no complications from homogeneous chemical reactions, then the current at an electrode will be determined by the transport and/or the electrode kinetics. Let us consider the steady-state current at an R.D.E. for the simple redox system:

$$Fe\,(CN)_6^{3-} \overset{e}{\underset{\leftarrow}{\rightleftharpoons}} Fe\,(CN)_6^{4-},$$

$$c_\infty^O \;=\; 3 \cdot 04 \, mM, \quad c_\infty^R \;=\; 1 \cdot 62 \, mM,$$

where c_∞ is the bulk concentration.

At the electrode surface

$$i \;=\; AF \left\{ k' e^{\dfrac{\alpha(E-E^\circ)F}{RT}} c_o^R - k' e^{\dfrac{-(1-\alpha)(E-E^\circ)F}{RT}} c_o^O \right\}, \quad (1.2)$$

where activity effects have been neglected, A is the area of the electrode, F is the Faraday, c_o^R and c_o^O are the concentrations of the species at the electrode surface, k' is a heterogeneous rate constant describing the electrode kinetics (measured in cm s^{-1}), E° is the standard electrode potential, α is the transfer coefficient,[2] and the exponential terms describe the effect of potential on the forward and backward rates.

Notice that if $i = 0$ then $c_o^R = c_\infty^R$, $c_o^O = c_\infty^O$ and

$$E - E^\circ \;=\; \frac{RT}{F} \ln \frac{c_\infty^O}{c_\infty^R},$$

which is the Nernst equation.

The potential difference at the electrode surface affects the forward and backward rate constants in different directions; they are equal at the standard electrode potential for the redox couple. Because of the exponential term a small difference in potential makes a large difference in rate. Another form of eqn (1.2) is

$$\iota \;=\; \frac{i}{A} \;=\; \iota_o^\circ \left\{ e^{\dfrac{\alpha(E-E^\circ)F}{RT}} c_o^R - e^{\dfrac{-(1-\alpha)(E-E^\circ)F}{RT}} c_o^O \right\}, \quad (1.3)$$

ι is a current density (current per unit area), and ι_o^o is the standard exchange current density. In practice the potential is usually measured not from E^o but from the equilibrium potential, E_E, at which $i = 0$ in the solution being studied. This is called the overpotential, η:

$$\eta = E - E_E.$$

Substitution in eqn (1.3) gives

$$\frac{i}{A} = \iota = \iota_o^o \left\{ e^{\frac{a\eta F}{RT}} c_o^R (c_o^O/c_\infty^R)^a - e^{\frac{-(1-a)\eta F}{RT}} c_o^O (c_\infty^R/c_\infty^O)^{1-a} \right\}. \quad (1.4)$$

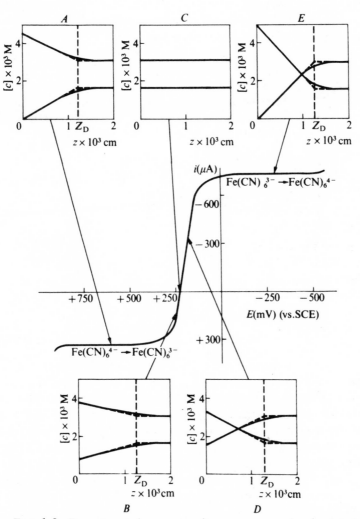

FIG. 1.2. Experimental current-voltage curves and calculated concentration profiles.

The effect of the concentration changes near the electrode is to alter c_o from c_∞.

Fig. 1.2 shows an experimental plot of the current as a function of potential together with the calculated concentration profiles close to the electrode. At point C the current is zero, the electrode is at its equilibrium potential, and the concentrations are uniform throughout the solution. At point A, on the other hand, the electrode is very positive and every Fe(II) species that reaches it is rapidly oxidized to Fe(III). As shown, the concentration of $Fe(CN)_6^{4-}$ is zero at the electrode surface and the current is determined wholly by the transport of $Fe(CN)_6^{4-}$ from the solution to the electrode. This current is called the *limiting current*. Conversely, at E the electrode is very negative, every Fe(III) is reduced, and the limiting current for the reduction of $Fe(CN_6)^{3-}$ is observed. When in this way the electrode is active enough to remove all the species reaching it, the potential of the electrode has no effect on the size of the transport-controlled limiting current. This is the reason for the characteristic plateaus in current voltage curves. B and D show concentration profiles for cases where the current is controlled partly by the electrode kinetics and partly by the transport.

Effect of Chemical Kinetics

The third factor mentioned above, the rate of homogeneous chemical change, may wholly or partly determine the current if there is an equilibrium between an electroactive species and a species that is inert on the electrode, with the latter in much larger concentration than the former.

A simple example of such a system is a weak acid buffer:

$$\text{solution } HA \rightleftharpoons H^+ + A^-,$$

$$\text{electrode } H^+ \xrightarrow{\ e\ } \tfrac{1}{2}H_2,$$

A platinum R.D.E. placed at ~ -1 V will reduce the H^+ to H_2 but not the HA, and for a weak acid $[HA] \gg [H^+]$. A plateau is observed on the current voltage curve at about -1 V. On this plateau the current is determined by the transport and by the homogeneous kinetics. Fig. 1.3 shows the concentration profiles for acetic acid. When the kinetics of the homogeneous system are fast, equilibrium between HA and H^+ is maintained throughout the diffusion layer and is only displaced very close to the electrode. In this 'reaction layer' the $[H^+]$ tends to zero on the electrode surface and the supply of H^+ is determined by the transport of HA and the kinetics of the equilibrium. The theoretical treatment of this problem was solved by Koutecky and Levich,[3] and

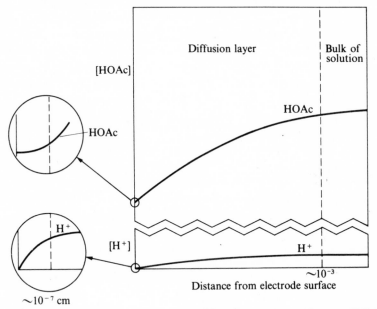

FIG. 1.3. Schematic concentration profiles for acetic acid at an R.D.E.

the rotating disc electrode has been successfully used to measure proton transfer to and from acetic and trimethyl acetic acids.[4] Caldin[5] has described the application of electrochemical measurements of this type.

References

1. ADAMS, R.N. (1969) In *Electrochemistry at solid electrodes*, Chap. 4, p. 67. Marcel Dekker.

2. VETTER, K.J. (1967) In *Electrochemical kinetics*. Academic Press.

3. KOUTECKY, J and LEVICH, V.G. (1956) *Zh. fiz. Khim.* **32**, 1565.

4. ALBERY, W.J. and BELL, R.P. (1963) *Proc. chem. Soc.* 169.

5. CALDIN, E.F. (1964) In *Fast reactions in solution*, Chap. 9, p. 164. Blackwells, Oxford.

2

THE SIMPLE ROTATING DISC ELECTRODE

The Hydrodynamics

IN the mathematical description of the rotating disc it is natural to use cylindrical polar coordinates as illustrated in Fig. 2.1: $r = 0$ at the

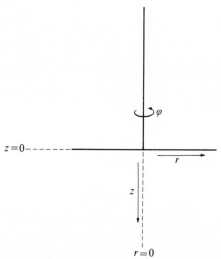

FIG. 2.1. Coordinates for the rotating disc system.

centre of the disc and $z = 0$ on the disc surface. Because of the cylindrical symmetry there are no variations in velocities, concentrations, etc. with ϕ, and all derivatives with respect to ϕ are equal to zero.

The first problem in describing the transport of species to an R.D.E. is to describe the pattern of flow imposed on the solution by rotating the disc. This problem was solved approximately by von Karman,[1] and the solution was improved by Cochran.[2] A recent review article by Riddiford[3] describes this work; here we shall merely give a qualitative picture.

The velocity of flow is resolved into three components, v_ϕ, v_r, and v_z, shown in Fig. 2.2. The fluid on the disc at $z = 0$ spins round

FIG. 2.2. Velocity components for the rotating disc system.

with it. Thus at $z = 0$, $v_\phi = r\omega$, where ω is the speed of rotation in rad s^{-1}, and $v_z = v_r = 0$. As z increases, the centrifugal force of the rotation produces shearing in the liquid driving it outwards from the centre. Thus, while v_r starts to rise, the rotational motion is damped out and v_ϕ falls. The liquid flung radially outwards can only be replaced by liquid flowing towards the disc from beneath it. The magnitude of v_z increases with z until it eventually reaches a limiting value in the bulk of the solution. Far away from the disc the radial and rotational motion are entirely damped out and v_r and v_ϕ are zero.

The mathematical solution of this problem[1,2] describes these velocities in terms of functions of the non-dimensional distance variable γ where

$$\gamma = \left(\frac{\omega}{\nu}\right)^{1/2} z = \frac{z}{Z_H}, \tag{2.1}$$

$$v_\phi = r\omega G(\gamma),$$

$$v_r = r\omega F(\gamma), \tag{2.2}$$

$$v_z = -(\omega\nu)^{1/2} H(\gamma), \tag{2.3}$$

ν is the kinematic viscosity in cm^2 s^{-1}, and Z_H is the thickness of the fluid boundary layer. Since the flow is towards the disc rather than away from it, v_z is negative. F, G, and $-H$ are plotted as functions of γ in Fig. 2.3. G describing the rotational motion starts at 1, (v_ϕ, at $z = 0$, $= r\omega$), and decreases as γ increases; F describing the centrifugal motion rises from zero to a maximum at a value of $\gamma \sim 1$, corresponding to $z = Z_H \sim 10^{-2}$ cm and then decreases again; $-H$ rises less steeply than F but carries on increasing to a limiting value of 0·885 as $\gamma \to \infty$. Notice that these functions depend only on γ and not on the radial distance. In particular, v_z describing the flow towards the disc is completely independent of r [see eqn (2.3)]. It is this feature of the hydrodynamics of the disc that makes the mathematical description of the transport to the disc much easier than that of other forced flow systems.

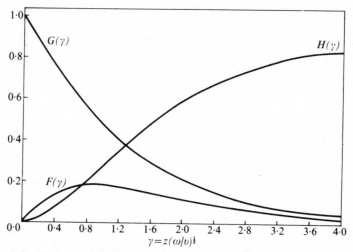

FIG. 2.3. Variation of the velocity components with the function γ at an R.D.E. $v_\phi = r\omega G(\gamma)$; $v_r = r\omega F(\gamma)$; $v_z = -(\omega/\nu)^{1/2}H(\gamma)$.

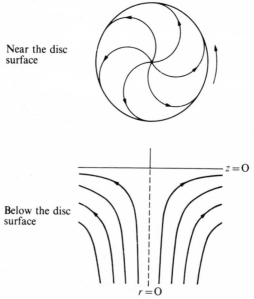

FIG. 2.4. Schematic streamlines for the rotating disc system.

To sum up, the disc acts as a pump, sucking solution towards it from the bulk and flinging it out centrifugally close to the disc surface.

Fig. 2.4 shows some schematic stream-lines for the rotating disc system.

The Basic Differential Equation for Transport

We now turn our attention to the transport of a species of concentration c from the bulk of the solution to the disc surface. The basic differential equation in cylindrical polar coordinates is

$$\frac{\partial c}{\partial t} = D\left(\frac{\partial^2 c}{\partial r^2} + \frac{1}{r}\frac{\partial c}{\partial r} + \frac{1}{r^2}\frac{\partial^2 c}{\partial \phi^2} + \frac{\partial^2 c}{\partial z^2}\right) -$$
$$- v_r \frac{\partial c}{\partial r} - \frac{v_\phi}{r}\frac{\partial c}{\partial \phi} - v_z \frac{\partial c}{\partial z}.$$

We can, for reasons of symmetry, immediately remove the derivatives with respect to ϕ to give

$$\frac{\partial c}{\partial t} = D\left(\frac{\partial^2 c}{\partial r^2} + \frac{1}{r}\frac{\partial c}{\partial r} + \frac{\partial^2 c}{\partial z^2}\right) - v_r \frac{\partial c}{\partial r} - v_z \frac{\partial c}{\partial z}. \qquad (2.4)$$

Thus in general, c, the concentration, is a function of time, the radial distance, and the distance normal to the electrode. A concentration change at any point (the left-hand side of the equation) can be brought about by diffusion, described by the first three terms on the right-hand side, or by convection, described by the last two terms. For the moment we have neglected the third transport process, the migration of an ion in an electric field. We have also assumed that D, the diffusion coefficient (dimensions $cm^2\ s^{-1}$), is a constant and is not a function of concentration or distance. Note that without the convection terms eqn (2.4) is Fick's second law of diffusion, and without the diffusion terms it is an equation of simple physical significance describing the effect of convection.

Many experiments are conducted under steady-state conditions; that is, sufficient time (1 or 2 s) is allowed for the diffusion and convection terms to balance each other and for $\partial c/\partial t$ to equal zero. A time-invariant pattern of concentration as a function of r and z is then set up in the solution described by

$$D\left(\frac{\partial^2 c}{\partial r^2} + \frac{1}{r}\frac{\partial c}{\partial r} + \frac{\partial^2 c}{\partial z^2}\right) = v_r \frac{\partial c}{\partial r} + v_z \frac{\partial c}{\partial z}. \qquad (2.5)$$

From the solution to the hydrodynamics[2] v_r and v_z may be replaced

by

$$v_r \simeq Crz \qquad (2.6)$$

and

$$v_z \simeq -Cz^2, \qquad (2.7)$$

where C, the convective constant, is given by

$$C = 0 \cdot 510 \, \omega^{3/2} \, \nu^{-1/2}. \qquad (2.8)$$

These equations are obtained from eqns (2.2) and (2.3) and the limiting forms of F and H as $\gamma \to 0$ (see Fig. 2.3). The reason why we can use the limiting form is that the main changes in concentration take place closer to the electrode ($\sim 10^{-3}$ cm) than the changes in the flow pattern, shown in Fig. 2.3.

The Boundary Conditions

To obtain the concentration pattern the differential equation (2.5) must be solved with four boundary conditions. These are

$$
\begin{aligned}
&\text{as } r \to \infty, && c \to c_\infty \, ; \\
&\text{as } r \to 0, && \partial c / \partial r \to 0; \qquad\qquad (2.9) \\
&\text{as } z \to \infty, && c \to c_\infty \, ; \\
&\text{as } z \to 0, && \text{the boundary condition appropriate} \\
& && \text{for the disc surface.}
\end{aligned}
$$

The first and third conditions state that far from the disc the concentration has its bulk value. The second condition is necessary since otherwise the $r^{-1} \partial c / \partial r$ term in eqn (2.5) goes to infinity. If $\partial c / \partial r$ is finite as $r \to 0$ there would be a cusp in a c vs. r plot, and this is physically impossible.

The Simplifications for a Disc Electrode

Although eqn (2.5) is the basic equation for ring-disc electrodes (R.R.D.E.), for the moment we will treat the simpler case of a disc electrode. The solution to this problem was first found by the distinguished Russian scientist V.G. Levich[4] in 1942. Since v_z does not depend on r [eqn (2.7)] there is uniform flow towards the electrode at any value of z, and since $\partial c / \partial r = 0$ at $r = 0$ there is no variation of c with r in the centre of the disc and $\partial c / \partial r$ will remain zero until the boundary condition on the disc surface changes with respect to r. Thus for a disc electrode of radius r_1, we can derive a solution for c, where c is uniform ($\partial c / \partial r = 0$) with respect to r for $r < r_1$, and where there is uniform current density (current per unit area) on the disc electrode. This means that the strip tease of eqn (2.5) can proceed further to give

$$D \frac{\partial^2 c}{\partial z^2} = v_z \frac{\partial c}{\partial z} = -Cz^2 \frac{\partial c}{\partial z} \qquad (2.10)$$

with the boundary conditions

$$z \to \infty, \qquad c \to c_\infty;$$

$$z = 0, \qquad j = D\left(\frac{\partial c}{\partial z}\right)_0 = k' c_0. \qquad (2.11)$$

The boundary condition at the disc electrode describes the flux j measured in mol $cm^{-2}s^{-1}$ in terms, first of the diffusion of the species out of the solution and secondly in terms of a heterogeneous reaction with a rate constant k' measured in $cm\ s^{-1}$ and c measured in mol cm^{-3}. The rate constant k' is similar to a homogeneous rate constant except first for its dimensions and secondly it will depend on the potential of the electrode.

The Mathematical Solution

Equation (2.6) is recast into non-dimensional form by writing

$$w = z C^{1/3} D^{-1/3}$$

to give

$$\frac{\partial^2 c}{\partial w^2} = -w^2 \frac{\partial c}{\partial w}.$$

Integration gives

$$\frac{\partial c}{\partial w} = \left(\frac{\partial c}{\partial w}\right)_0 \exp\left(-\tfrac{1}{3} w^3\right)$$

and

$$c = c_0 + \left(\frac{\partial c}{\partial w}\right)_0 \int_0^w \exp\left(-\tfrac{1}{3} w^3\right) dw, \qquad (2.12)$$

from which, as $w \to \infty$,

$$c_\infty = c_0 + \left(\frac{\partial c}{\partial w}\right)_0 3^{1/3} \Gamma(4/3). \qquad (2.13)$$

The constant $3^{1/3} \Gamma(4/3) = 1 \cdot 288$ and will be written as A_1.
Equation (2.11) gives

$$\left(\frac{\partial c}{\partial w}\right)_0 = k' C^{-1/3} D^{-2/3} c_0$$

and substitution in eqn (2.13) gives

$$j = \frac{k' c_\infty}{1 + A_1 k' C^{-1/3} D^{-2/3}}.$$

The current is given by

$$i = \pi r_1^2 n F j.$$

As the potential of the electrode is varied k' changes from a very small value, where

$$A_1 k' C^{-1/3} D^{-2/3} \ll 1 \quad \text{and} \quad j = k' c_\infty,$$

to a much larger value where

$$A_1 k' C^{-1/3} D^{-2/3} \gg 1 \quad \text{and} \quad j_L = \frac{Dc_\infty}{A_1 D^{1/3} C^{-1/3}} = \frac{Dc_\infty}{Z_D}. \quad (2.14)$$

This change is the mathematical description of the change discussed in Chapter 1, from the current being controlled by the electrode kinetics (k') to the current being controlled by the transport (D/Z_D). Z_D has the dimensions of a length and describes the thickness of the diffusion layer. Fig. 2.5 shows a plot of eqn (2.12), giving a concentration profile for the transport limited current where $c_o = 0$. Z_D divides

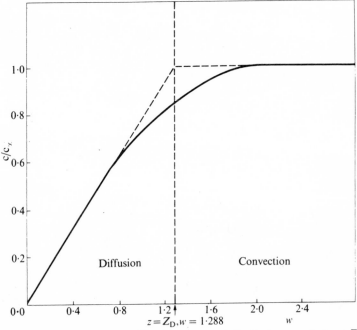

FIG. 2.5. Concentration profile for transport limited current at an R.D.E.

the regions of the solution into the well-stirred region, where convection is predominant, and the stagnant layer, where diffusion is the main transport process. When

$$z = Z_D, \quad w = A_1 = 1.288.$$

Substituting from eqn (2.8),

$$Z_D = A_1 D^{1/3} C^{-1/3} = 0.643\, W^{-1/2} \nu^{1/6} D^{1/3}, \qquad (2.15)$$

where W is now measured in Hz (the more practical unit). For $D = 10^{-5}\,\text{cm}^2\text{s}^{-1}$, $\nu = 10^{-2}\,\text{cm}^2\text{s}^{-1}$ and $W = 40\,\text{Hz}$, $Z_D = 10^{-3}\,\text{cm}$, from eqns (2.1) and (2.15)

$$\frac{Z_D}{Z_H} = \left(\frac{D}{\nu}\right)^{1/3} \simeq 10^{-1}.$$

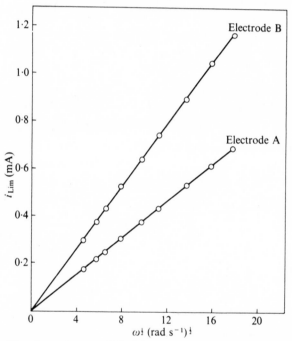

FIG. 2.6. The dependence of limiting currents on rotation speed. $[\text{KI}] = 1.05\,\text{mM}$ $[\text{H}_2\text{SO}_4] = 0.10\,\text{M}$. Radius r_1: Electrode A 0.3635 cm; Electrode B 0.4770 cm.

This ratio is the reason why we need only take the limiting expressions for v_z and v_r. In very accurate work involving the mass transport of iodine to a zinc disc,[5] Gregory and Riddiford showed that better agreement between theory and experiment could be obtained if a second term was included in the expression for v_z. However, the correction was only a few per cent and it is likely that in electrochemical measurements the corrections for the migration of the ions[6] will be just as significant.

Combination of eqns (2.14) and (2.15) gives

$$j_L = 1\cdot554\ D^{2/3}\ \nu^{-1/6}\ W^{1/2}c_\infty. \tag{2.16}$$

This equation predicts that the limiting current should be proportional to the square root of the rotation speed. Many workers[3] have shown that this is true. Fig. 2.6 shows some typical results.[7] Further applications of the basic R.D.E. are discussed in Riddiford's useful review article.[3]

References

1. VON KÁRMÁN, T. (1921) Z. angew. Math. Mech. 1, 233.

2. COCHRAN, W.G. (1934) Proc. Camb. Phil. Soc. math. phys. Sci. 30, 365.

3. RIDDIFORD, A.C. (1966) In Advances in electrochemistry and electrochemical engineering (ed. Delahay), Vol. 4. Interscience.

4. LEVICH, V.G. (1942) Acta phys.-chim. URSS 17, 257.

5. GREGORY, D.P. and RIDDIFORD, A.C. (1956) J. chem. Soc. 3756.

6. ALBERY, W.J. (1965) Trans. Faraday Soc. 61, 2063.

7. HITCHMAN, M.L. (1968) D. Phil. Thesis, Oxford.

3
THE COLLECTION EFFICIENCY

The Ring-Disc Electrode

FIG. 3.1 shows a ring-disc electrode (R.R.D.E.). This electrode was
first developed by Frumkin and Nekrasov[1] to detect unstable intermediates in electrode reactions. The electrode consists of a central
disc electrode surrounded by a concentric insulating annulus and then

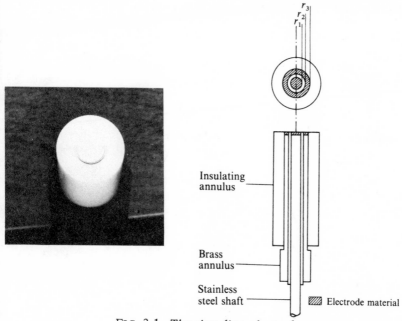

FIG. 3.1. The ring-disc electrode.

a concentric ring electrode. Thus we now have two electrodes at which
experimental investigations can be made and the potentials and currents
at each electrode can be controlled separately from each other. This
allows, for instance, the use of the ring electrode to provide extra
information about processes occurring on the disc electrode, or the

use of the disc electrode as a generator and the ring electrode as a detector of unstable species.

Definition of N_o

In order to do this in a quantitative way we must be able to describe the transport of species from the disc electrode to the ring. Thus the first and simplest problem is to calculate the collection efficiency, N_o, of the electrode. An approximate solution to this problem was first found by Ivanov and Levich[2] and a more exact analytical expression was derived by Albery and Bruckenstein.[3,4]

Imagine that the disc electrode is set to pass a constant current, which generates a constant flux of an 'intermediate', and the ring electrode is set at a potential such that all the intermediate that reaches it is destroyed. That is, the surface concentration of the intermediate on the ring electrode is zero. Some of the intermediate will escape into the bulk of the solution and so the ring current will be a fraction (less than one) of the disc current. This fraction is called the *collection efficiency*.

Thus we have

$$\text{disc } A \rightarrow B;$$

$$\text{ring } B \rightarrow A;$$

$$N_o = -i_R / i_D.$$

The negative sign arises from the fact that the currents pass in the opposite directions. Table 3.1 lists some typical A/B systems.

TABLE 3.1

Typical systems for measuring the collection efficiency of an R.R.D.E.

Disc	Ring	Reference
$Br^- \xrightarrow{-e} \frac{1}{2}Br_2$	$\frac{1}{2}Br_2 \xrightarrow{e} Br^-$	4
$Ag \xrightarrow{-e} Ag^+$	$Ag^+ \xrightarrow{e} Ag$	4
$Fe(CN)_6^{4-} \xrightarrow{-e} Fe(CN)_6^{3-}$	$Fe(CN)_6^{3-} \xrightarrow{e} Fe(CN)_6^{4-}$	5
$Cu^{2+} \xrightarrow{e} Cu^+$	$Cu^+ \xrightarrow{-e} Cu^{2+}$	4
$Fe^{3+} \xrightarrow{e} Fe^{2+}$	$Fe^{2+} \xrightarrow{-e} Fe^{3+}$	6
$O_2 \xrightarrow{2e} H_2O_2$	$H_2O_2 \xrightarrow{-2e} O_2$	1
$O{=}\langle\text{ring}\rangle{=}O \xrightarrow{-2e} HO{-}\langle\text{ring}\rangle{-}OH$	$HO{-}\langle\text{ring}\rangle{-}OH \xrightarrow{-2e} O{=}\langle\text{ring}\rangle{=}O$	1,7

Differential Equation and Boundary Conditions

Although starting from the centre of the electrode system the concentrations are uniform with respect to r below the disc electrode, at the outside edge of the disc electrode there is a discontinuity in the concentration gradients leading to a variation of the concentration of B with r as it falls away to zero both on the ring electrode and a long way from the disc. Fig. 3.2 shows a schematic contour diagram

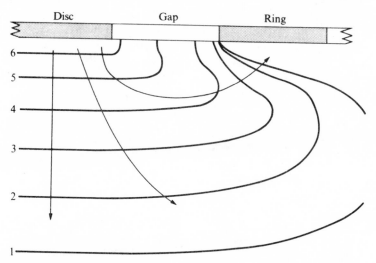

FIG. 3.2. Schematic concentration profile at a ring-disc electrode.

of concentration against r and z. B is transported away from the disc electrode to the ring electrode by diffusing into the solution, being swept outwards by the radial convection and then diffusing back on to the ring electrode.

The problem of this transport is solved by dividing the solution into three zones.[2] An R.R.D.E. is characterized by three radii:

r_1, radius of disc electrode;

r_2, inner radius of ring electrode;

r_3, outer radius of ring electrode.

The first zone is the region of the disc electrode, $r < r_1$; the second zone is the zone of the insulating gap, $r_1 < r < r_2$; and the third zone is the region of the ring electrode, $r_2 < r < r_3$. Each zone has its own boundary condition on the surface of the disc.

The differential equation to be solved is eqn (2.5):

$$D\left(\frac{\partial^2 b}{\partial r^2} + \frac{1}{r}\frac{\partial b}{\partial r} + \frac{\partial^2 b}{\partial z^2}\right) = v_r\frac{\partial b}{\partial r} + v_z\frac{\partial b}{\partial z}, \qquad (2.5)$$

B

with the boundary conditions:

all zones $\qquad\qquad\qquad\qquad z \to \infty, \quad b \to 0;$ $\qquad$ (3.1)

zone of the disc $\qquad\quad r < r_1, \quad r \to 0, \quad \partial b/\partial r \to 0;$ $\qquad$ (3.2)
electrode

$$z = 0, \quad \left(\frac{\partial b}{\partial z}\right)_0 = -\frac{i_D}{\pi r_1^2 nFD};$$ $\qquad$ (3.3)

zone of the gap $\qquad\quad r_1 < r < r_2, \quad z = 0, \quad \left(\frac{\partial b}{\partial z}\right)_0 = 0;$ $\quad$ (3.4)

zone of the ring $\qquad\quad r_2 < r < r_3, \quad z = 0, \quad b = 0.$ $\qquad$ (3.5)
electrode

The first boundary condition states that the bulk concentration of the electrogenerated intermediate is zero; the second condition has been discussed in Chapter 2; the third describes the uniform disc current on the disc electrode; the fourth the fact that nothing can react on the insulating surface between the disc and the ring; and the fifth is the condition for the limiting current at the ring electrode.

Mathematical Solution

In the zone of the disc electrode we have the same solution as that discussed in Chapter 2; the contours in Fig. 3.2 run parallel to the disc surface in this region since $\partial b/\partial r$ is zero. In the region of the gap they are perpendicular to the disc surface [eqn (3.4)] and they run down to zero both on the ring electrode and as z and $r \to \infty$.

The mathematical soltion for the region of the disc at $r = r_1$ provides a boundary condition at $r = r_1$ for the region of the gap. Similarly, when the concentration profile in the gap has been solved, the value of b at $r = r_2$ obtained from the gap provides a boundary condition for the ring. The full eqn (2.5) requires two radial boundary conditions for each zone, but fortunately we can show that radial diffusion is negligible compared to radial convection. Taking typical values, $\nu = 10^{-2} \text{ cm}^2 \text{ s}^{-1}$, $\omega = 2 \times 10^2 \text{ rad s}^{-1}$, $r = 0\cdot5 \text{ cm}$, and $D = 10^{-5} \text{ cm}^2 \text{ s}^{-1}$, from eqn (2.5)

$$\frac{D}{r v_r} = \frac{3 \times 10^{-9}}{z}.$$

Therefore for values of z greater than $10\,\text{Å}$ the radial diffusion terms are much smaller than the radial convection. The reason for this is that the radial velocity depends upon r. Outside the disc electrode r is usually greater than several millimetres. This is to be contrasted with the typical values of $z \sim 10^{-3}$ cm; from eqns (2.6) and (2.7):

$$|v_r/v_z| = r/z.$$

,Hence close to the electrode the main transport processes are radial convection and diffusion normal to the electrode. Equation (2.5) then becomes

$$D \frac{\partial^2 b}{\partial z^2} = v_r \frac{\partial b}{\partial r} + v_z \frac{\partial b}{\partial z}. \tag{3.6}$$

This equation can be solved analytically with the boundary conditions (3.1) to (3.5) to give the ring current

$$i_R = nFD\, 2\pi \int_{r_2}^{r_3} \left(\frac{\partial b}{\partial z}\right)_0 r\, dr$$

and hence

$$N_o = -i_R/i_D.$$

The method of solution is to express the equation in non-dimensional variables:

$$w = (C/D)^{1/3} z,$$

$$u = \frac{\pi r_1^2 \, n\, F\, D^{2/3}\, C^{1/3}}{|i_D|} b, \tag{3.7}$$

$$\xi_1 = \tfrac{1}{3}\{(r/r_1)^3 - 1\},$$

$$\xi_2 = \tfrac{1}{3}\{(r/r_1)^3 - (r_2/r_1)^3\},$$

$$y = (r/r_1)w.$$

The function u is always chosen so that

$$z \to \infty, \qquad u \to 0;$$

$$r < r_1, \qquad (\partial u/\partial w)_0 = -1;$$

$$r_1 < r < r_2, \qquad (\partial u/\partial w)_0 = 0. \tag{3.8}$$

In addition we have for this particular problem

$$r_2 < r < r_3, \qquad u = 0. \tag{3.9}$$

The differential equation (3.6) becomes

$$\frac{\partial u}{\partial \xi_n} = \frac{1}{y}\frac{\partial^2 u}{\partial y^2},$$

where $n = 1$ for the gap and $n = 2$ for the ring.

The Laplace transform of this equation with respect to ξ_n has the general form

$$y\, s_n\, \bar{u} - y u_{\xi_n = 0} = \frac{\partial^2 \bar{u}}{\partial y^2}.$$

Solutions[3] in terms of Airy functions[8] can be found for this equation.

Another advantage of working with Laplace transforms is that we do not have to solve the complete description of b with r and z, but one can just operate with the equations and the boundary conditions. Indeed, after transforming with respect to ξ_1 for the zone of the gap, we do not even invert to obtain u at $r = r_2$ but we carry on with the second transformation with respect to ξ_2 for the zone of the ring. Then:[4]

$$N_0 = 2\,\mathcal{L}_2^{-1}\,\mathcal{L}_1^{-1}\left\{\frac{1}{s_2}\left(\frac{\partial \bar{\bar{u}}}{\partial y}\right)_{y=0}\right\}_{\substack{\xi_1 = \xi_1' \\ \xi_2 = \xi_2'}}$$

where

$$\xi_1' = \tfrac{1}{3}\alpha = \tfrac{1}{3}\{(r_2/r_1)^3 - 1\}, \tag{3.10}$$

and

$$\xi_2' = \tfrac{1}{3}\beta = \tfrac{1}{3}\{(r_3/r_1)^3 - (r_2/r_1)^3\}. \tag{3.11}$$

The double transform can be inverted to obtain the analytical solution in terms of the two geometrical parameters α and β:

$$N_0 = 1 - F(\alpha/\beta) + \beta^{2/3}\{1 - F(\alpha)\} -$$
$$(1 + \alpha + \beta)^{2/3}[1 - F\{(\alpha/\beta)(1 + \alpha + \beta)\}] \tag{3.12}$$

where

$$F(\theta) \equiv \frac{3^{1/2}}{4\pi}\ln\left\{\frac{(1 + \theta^{1/3})^3}{1 + \theta}\right\} + \frac{3}{2\pi}\cdot\arctan\left(\frac{2\theta^{1/3} - 1}{3^{1/2}}\right) + \frac{1}{4}. \tag{3.13}$$

This function is tabulated in Appendix 1 for values of $\theta < 1$. For $\theta > 1$,

$$F(\theta) = 1 - G(\phi), \tag{3.14}$$

where $\phi = 1/\theta$. Values of $G(\phi)$ for $\phi < 1$ $(\theta > 1)$ are given in Appendix 1. There is also a table of values of N_0 for common radius ratios. Table 3.2 compares observed and theoretical values of N_0 for electrodes of very different geometry. Very good agreement between theory and experiment is found.

Effect of Bulk Concentration of Intermediate

In the problem discussed so far we have assumed that the bulk concentration of the intermediate is zero. We now extend the solution to describe what happens when the bulk concentration of the intermediate

ATION

TABLE 3.2

Comparison of calculated and experimental values of N_o

	A	B	C	D	E
r_1 (cm)	0·3869	0·4769	0·3480	0·3672	0·3635
r_2 (cm)	0·3981	0·4869	0·3860	0·3763	0·3779
r_3 (cm)	0·4051	0·5221	0·4375	0·4369	0·4839
α	0·0894	0·064	0·3647	0·0762	0·122
β	0·1507	0·249	0·6223	0·6082	1·235
Calculated N_o	0·090$_4$	0·214	0·261	0·318	0·402
Observed N_o	0·097$_4$	0·215	0·263	0·328	0·404

is not zero. When no disc current is flowing there will be a finite ring current. The problem of the limiting current at a ring electrode was first solved by Levich.[9] The same solution is obtained by the double Laplace transform method and may be simply expressed [9] in the form

$$i^o_{R,L} = \beta^{2/3} i_{D,L},\qquad(3.15)$$

where $i^o_{R,L}$ is the limiting current on the ring electrode for $i_D = 0$, $i_{D,L}$ is the limiting current on the disc electrode, and β is a geometric function of the radii r_1, r_2, and r_3 defined in eqn (3.11).

It is interesting to point out that β can be easily measured from the ratio of the two limiting currents. Since N_o is a function of only α and β, the geometry of the electrode can be found from two electrochemical measurements and a value of r_1. This procedure is about as accurate as measuring the radii with a travelling microscope for electrodes of the geometry in Table 3.2. However, for ultra-thin-gap, thin-ring electrodes it may be the best approach.

Returning to the effect of the bulk concentration on the ring current, instead of the previous definition of u, eqn (3.7), we write

$$u = \frac{\pi r_1^2 nFD^{2/3} C^{1/3}}{|i_D|} (b - b^o),$$

where b^o is the concentration distribution when $i_D = 0$. The boundary conditions for b^o are

$$z \to \infty, \quad b^o \to b_\infty;$$

$$r < r_1, \quad (\partial b/\partial z)_o = 0 \quad \text{(no disc current)};$$

$$r_1 < r < r_2, \quad (\partial b/\partial z)_o = 0;$$

$$r_2 < r < r_3, \quad b_o^o = 0;$$

and thus the boundary conditions for u are

$$z \to \infty, \quad u \to 0;$$

$$r < r_1, \quad (\partial u/\partial w)_o = -1;$$

$$r_1 < r < r_2, \quad (\partial u/\partial w)_o = 0;$$

$$r_2 < r < r_3, \quad u = 0.$$

These are exactly the same boundary conditions for u as in eqns (3.8) and (3.9). It obeys the same differential equation and so the same solution applies.

Hence one obtains directly [10] that

$$i_{R,L} = i_{R,L}^o - N_o i_D,$$

where $i_{R,L}^o$ is the limiting ring current when $i_D = 0$. This simple equation states that if one generates an intermediate on the disc then the current for its destruction on the ring will be increased by $N_o i_D$ from the background current of $i_{R,L}^o$. On the other hand, if we destroy the intermediate on the disc then the ring current will be decreased by $N_o i_D$ because the disc electrode is removing the electroactive species.

Shielding Effect

If the disc is at such a potential that the limiting current is being passed then, from eqn (3.15),

$$i_{D,L} = \beta^{-2/3} i_{R,L}^o$$

and

$$i_{R,L} = i_{R,L}^o (1 - N_o \beta^{-2/3}).$$

It can be shown that $N_o \beta^{-2/3} < 1$ and therefore the factor $(1 - N_o \beta^{-2/3})$ is always positive. This factor is called the *shielding factor*, and describes the maximum reduction in the ring current that can be achieved by the disc electrode. Its importance may be in analytical applications of the ring-disc electrode. The disc electrode would be used to remove species A which might interfere with the electrochemical determination of species B on the ring electrode. Thin-ring, thin-gap electrodes where $N \to \beta^{2/3}$ have low shielding factors and would therefore have the preferred geometry for this type of application.

Current-Voltage Curve at the Ring Electrode

This type of analysis of the effect of the disc electrode on the ring current can be extended further [10] to a reversible redox couple:

$$A \pm ne \rightleftarrows B,$$

If we assume that A and B have the same diffusion coefficients then over the whole disc surface

$$\frac{\partial (a + b)}{\partial z} = 0, \tag{3.16}$$

since on either electrode

$$\left(\frac{\partial a}{\partial z}\right)_r = -\left(\frac{\partial b}{\partial z}\right)_r.$$

Equation (3.16) means that $a + b$ is a constant and $a + b = a_\infty + b_\infty$, Also, if the system is reversible on the ring electrode,

$$a_o / b_o = K_R.$$

Thus

$$a_o = \frac{K_R (a_\infty + b_\infty)}{1 + K_R}$$

and

$$b_o = \frac{(a_\infty + b_\infty)}{1 + K_R}.$$

Thus a_o and b_o depend on the bulk concentrations and K_R which is fixed by the ring potential; they do not depend on the disc current. Now we define

$$u = \frac{\pi r_1^2 nFD^{2/3} C^{1/3}}{|i_D|} (a - a^o)$$

where a^o is the concentration distribution of a when $i_D = 0$. Then u will obey the same boundary conditions as in eqns (3.8) and (3.9), since both a and a^o tend to a_∞ in the bulk of the solution, and both equal a_o on the ring electrode. Thus once again

$$i_R = i_R^o - N_o i_D.$$

Fig. 3.3 shows the effect of the disc current on the current voltage curve at the ring electrode. A more detailed discussion is given in Ref. 10, where the assumption of equal diffusion coefficients is not made.

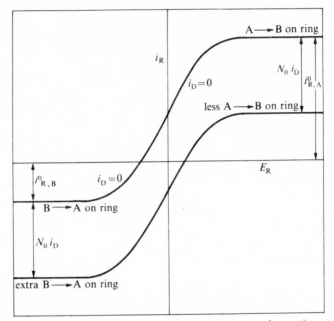

F_IG_. 3.3. Current-voltage curves at a ring electrode.

Numerical Method of Solution

Recently, numerical methods have been applied to the calculation of concentration profiles around a ring-disc electrode. Feldberg's digital simulation technique[11] has been adapted by Bard and Prater[12,13] to the ring-disc electrode. Feldberg has described his technique in some detail.[14]

Briefly, the region below the electrode is divided into a number of 'boxes' in both the r and z directions (see Fig. 3.4). J describes the

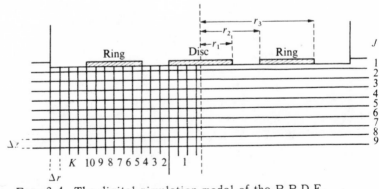

F_IG_. 3.4. The digital simulation model of the R.R.D.E.
(From Ref. 12, by permission of the Electrochemical Society.)

boxes in the normal direction and K the boxes in the radial direction. The concentration of each box is the concentration at its centre. The calculation starts by assuming that each box has the bulk concentration, c_∞. The electrode is then switched on and the values in the $J = 1$ (the electrode surface) and $J = 2$, the layer next to the electrode, are adjusted with the appropriate boundary conditions on the disc surface for the generation or discharge of material over a small time Δt. This gives rise to concentration gradients. The effect of diffusion normal to the electrode over the time Δt is next calculated by using a finite difference representation of Fick's second law, and the values in all the boxes are then adjusted for the effects of this type of transport. Then a similar treatment is applied for the transport by convection both in a radial and a normal direction for a time Δt, and again the concentrations are all adjusted. Meanwhile, back at the ranch $(z = 0)$, material is being discharged or generated on the electrodes and this is again calculated over a period of time Δt. The iterations continue until a steady state is reached. Even the University of Texas does not have unlimited funds for computing and Bard and Prater have developed a programme that speeds up the iterative process.

Their results for N_o are compared with the analytical theory in Table 3.3. Each calculation took about 20 s. Everyone can be pleased that good agreement is obtained.

TABLE 3.3

Comparison of values of N_o obtained analytically and numerically

Radius ratios		Collection efficiencies, N_o	
r_2/r_1	r_3/r_1	Analytical solution[4]	Numerical solution[12]
1·05	1·261	0·340	0·339
1·05	1·361	0·409	0·408
1·05	1·472	0·464	0·463
1·07	1·161	0·210	0·209
1·07	1·271	0·323	0·321
1·07	1·371	0·391	0·391
1·07	1·483	0·449	0·449
1·09	1·201	0·226	0·226
1·09	1·301	0·321	0·320
1·09	1·521	0·447	0·449

References

1. FRUMKIN, A.N. and NEKRASOV, L.I. (1959) *Dokl. Akad. Nauk SSSR* 126, 115.

2. IVANOV, Y.B. and LEVICH V.G. (1959) *Dokl. Akad. Nauk SSSR* 126, 1029.

3. ALBERY, W.J. (1966) *Trans. Faraday Soc.* 62, 1915.

4. ALBERY, W.J. and BRUCKENSTEIN, S. (1966) *Trans. Faraday Soc.* 62, 1920.

5. HITCHMAN, M.L. (1968) D. Phil. Thesis, Oxford.

6. ALBERY, W.J., HITCHMAN, M.L., and ULSTRUP, J. (1968) *Trans. Faraday Soc.* 64, 2831.

7. ADAMS, R.N. (1969) In *Electrochemistry at solid electrodes*, Chap. 4, p. 99. Marcel Dekker.

8. *Handbook of mathematical functions* (1965) (eds. M. Abramowitz and I.A. Stegun), Chap. 10, pp. 446, 447. Dover.

9. LEVICH, V.G. (1962) In *Physicochemical hydrodynamics*, pp. 107, 314. Prentice Hall.

10. ALBERY, W.J., BRUCKENSTEIN, S., and NAPP, D.T. (1966) *Trans. Faraday Soc.* 62, 1932.

11. FELDBERG, S.W. and AUERBACH, C. (1964) *Analyt Chem.* 36, 505.

12. BARD, A.J. and PRATER, K.B. (1970) *J. electrochem. Soc.* 117, 207.

13. BARD, A.J. and PRATER, K.B. (1970) *J. electrochem. Soc.* 117, 335.

14. FELDBERG, S.W. (1969) In *Electroanalytical chemistry* (ed. A.J. Bard), Vol. 3, pp. 199–296. Marcel Dekker.

4

NON-UNIFORMITY OF CURRENT DISTRIBUTION ON THE DISC ELECTRODE

Introduction

ALTHOUGH from considerations of the transport and surface boundary conditions the concentration profiles for the disc electrode are independent of the radial distance, Newman[1,2] has recently shown that when the effects of the electric field are taken into account the simple solution may break down. The reason for this is that the surface boundary condition on the disc electrode is dependent on the potential difference between the metal electrode and the solution close to the electrode. The potential of the metal electrode is certainly uniform, and until recently it had been assumed that the potential of the solution close to the electrode also did not vary with r. However, the passage of current in the bulk of the solution outside the diffusion layer destroys the radial uniformity.

Primary Current Distribution

Newman[1] first calculated the primary current distribution and its associated potential distribution in the bulk of the solution. This is shown in Fig. 4.1. This is the current distribution we would observe for the case where the potential in the solution close to the electrode is uniform and the counter electrode is a very long way from the disc electrode. While the contours of equal potential run almost parallel to the surface of the disc electrode when close to the electrode, further out from the electrode they become hemispherical in shape. The distance over which the effect operates is related directly to the radius of the electrode. Thus the distance in the z direction in Fig. 4.1 is much larger than the thickness of the diffusion layer. The primary current distribution concentrates the current on the outside edge of the disc electrode owing to the fact that the current flow reaches the electrode not just from directly below the electrode but also from

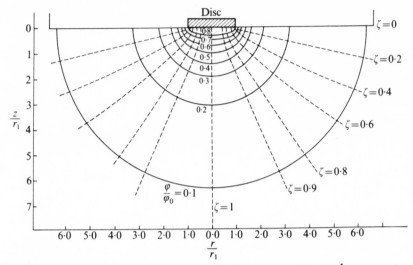

FIG. 4.1. Current and potential distribution near an R.D.E.[1] Lines of equal potential——; lines of current flow $---$. $z/r_1 = \xi\zeta$; $r/r_1 = \sqrt{(1 + \xi^2)(1 - \zeta^2)}$; $r_1 =$ radius of disc electrode; $\phi/\phi_0 = 1 - (2/\pi)$ $\tan^{-1}\xi$; $\phi =$ potential of disc electrode. (From Ref. 1, by permission of the Electrochemical Society).

outside it. The fact that an assumption of a uniform distribution of potential does not lead to a uniform distribution of current means, in general, that potential, current, and concentration distributions may vary with r for the disc electrode. Newman has presented numerical solutions [2] to the problem of what happens when these distributions are controlled by a mixture of electrode kinetics, concentration polarization, and ohmic drop in the bulk of the solution. We will, however, present a more approximate treatment in terms of resistances.

Effect of Charge Transfer Resistance

We may imagine the difference between the centre and the outside of the disc electrode to be represented by the resistances shown in Fig. 4.2. The small areas of the electrode surface have been chosen so that they are equal. R_{CT} represents the charge transfer resistance for each small area of electrode. It is not a real resistance but it has the dimensions of resistance and is defined as

$$d(1/R_{CT}) = d\pi\, r\, d\, r \frac{\partial \iota}{\partial \eta},$$

where η is the overpotential and ι is the current per unit area when the current is controlled entirely by the electrode kinetics. R_Ω and R'_Ω are the resistance to current flow in the bulk of the solution; they cannot really be separated in this way but the extra cross-sectional

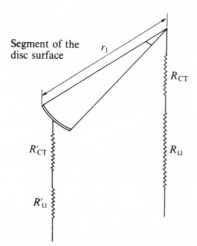

FIG. 4.2. Resistances to current flow at an R.D.E.

area available for the outside of the disc means that $R'_\Omega < R_\Omega$.
Overall, this resistance may be averaged to give $\bar{R}_\Omega$, which is defined[1]
as follows:

$$\bar{R}_\Omega = \frac{1}{4 \kappa_\infty r_1},$$

where κ_∞ is the specific conductivity of the bulk of the solution. $\bar{R}_\Omega$
is the effective resistance of the bulk of the solution when we have
the primary current distribution.

Similarly, $\bar{R}_{CT}$, the average charge transfer resistance, may be
calculated for a simple redox system from eqn (1.3) as if the different
elements were resistances in parallel:

$$\frac{1}{\bar{R}_{CT}} = \int_0^{r_1} \frac{2\pi r\, dr}{(\partial \eta/\partial \iota)}$$

$$= \frac{2\pi F \iota_o^o (c_\infty^O)^\alpha (c_\infty^R)^{1-\alpha}}{RT} \times$$

$$\times \int_0^{r_1} \left\{ \alpha e^{\frac{\alpha \eta F}{RT}} + (1-\alpha) e^{\frac{-(1-\alpha)\eta F}{RT}} \right\} r\, dr,$$

where for the moment concentration polarization has been ignored. The
terms in η must be kept inside the integral since η is a function of r.

Two limiting cases can be considered. First, when $\eta F/RT$ is
small, that is, for small displacements of potential from the equilibrium
potential; and secondly, when $|\eta F/RT| > 3$ and only one term in eqn
(1.3) need be considered.

In the first case the exponentials are assumed to be 1 and

$$\frac{1}{\bar{R}_{CT}} = \frac{\pi r_1^2 F \iota_o^o (c_\infty^O)^\alpha (c_\infty^R)^{1-\alpha}}{RT} .$$ (4.1)

In the second case (taking $\eta + ve$)

$$\frac{1}{\bar{R}_{CT}} = \frac{\alpha F \iota_{Av} \pi r_1^2}{RT} .$$ (4.2)

The non-uniformity of the current distribution can be considered in terms of the dimensionless parameter

$$\rho = \frac{\bar{R}_\Omega}{\bar{R}_{CT}} .$$

If $\bar{R}_{CT}$ is determined by eqn (4.1) then this parameter is equivalent to Newman's parameter[2] J:

$$\rho = \frac{\pi J}{4} ,$$

and if $\bar{R}_{CT}$ is determined by eqn (4.2) then the parameter is equivalent to Newman's parameter $\beta\delta$:

$$\rho = \frac{\pi}{4} \beta\delta.$$

Fig. 4.3 shows results from Newman's numerical calculations[2] in terms of the new variable ρ. The y-axis shows the ratio of the current density at any value of r/r_1 to the average current density. As can be seen, this treatment in terms of resistances unifies the two cases considered by Newman. When ρ is small $\bar{R}_\Omega$ is much less than $\bar{R}_{CT}$. However, since the non-uniformity arises from the variation with r of R_Ω (Fig. 4.2), when it is swamped by $\bar{R}_{CT}$ it will have little effect and when $\rho < 0.1$ the effect is negligible. On the other hand, when ρ tends to infinity we obtain the primary current distribution.

Effect of Concentration Polarization

So far this discussion has ignored concentration polarization. Newman has treated this numerically.[2] However it can be considered approximately in much the same way as the charge transfer resistance, since like the electrode reactions the concentration polarization occurs at much shorter distances than the ohmic drop. Thus R_{CT} in Fig. 4.2 is replaced by R_{E1}, where this is the electrode resistance due both to charge transfer and to concentration polarization. It could also include

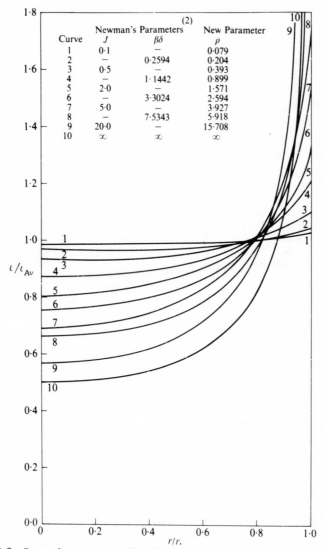

FIG. 4.3. Secondary current distribution at an R.D.E. (From Ref. 2,
by permission of the Electrochemical Society.)

any other features of the system that control the current close to the
electrode, for instance homogeneous kinetics.

We write

$$\bar{R}_{E1} + \bar{R}_{\Omega} = \frac{d\eta}{di},$$

where $d\eta/di$ is the observed slope of the current voltage curve.

From

$$\rho = \frac{\bar{R}_\Omega}{\bar{R}_{El}}$$

and

$$\bar{R}_\Omega = \frac{1}{4 r_1 \kappa_\infty}$$

we obtain

$$\rho = \frac{1}{\dfrac{4 r_1 \kappa_\infty}{(d i/d\eta)} - 1}. \tag{4.3}$$

Fig. 4.3 associates a particular radial current distribution with a particular value of ρ. Newman's detailed calculations [2] show that this is not strictly true and that, for instance, concentration polarization may introduce a maximum in the ι vs. r curve. However, Fig. 4.3 is a good approximate guide to the degree of non-uniformity. Values of the current density at $r = 0$ divided by the average current density calculated from ρ and Fig. 4.3 are generally in good agreement with the values calculated by Newman; a typical comparison is shown in Fig. 4.4.

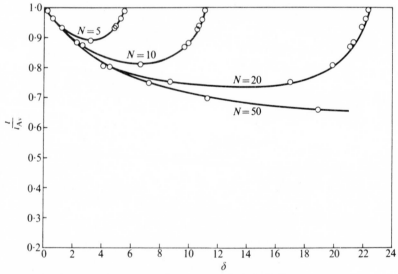

FIG. 4.4. Current density at the centre of an R.D.E.

$$\delta = \iota_{Av} \frac{zF}{RT} \frac{r_1}{\kappa_\infty}; \qquad N = \iota_L \Gamma\left(\frac{4}{3}\right) \frac{zF}{RT} \frac{r_1}{\kappa_\infty},$$

where ι_L = limiting current density. Curves taken from Fig. 8 of Ref. 2. Points: ι/ι_{Av} values obtained from Fig. 4.3 for given ρ value, and δ values calculated from eqn (4.3) and eqns (21), (27), and (37) of Ref. 2. (From Ref. 2, by permission of the Electrochemical Society.)

The curves are Newman's results, while the points are values calculated by this approximate treatment. The advantage of this approach is that we have only to calculate the one parameter ρ from the three easily obtained quantities, r_1, κ_∞, and $\mathrm{d}i/\mathrm{d}\eta$.

Condition for Uniform Current Distribution

To avoid the complications of non-uniform current distribution on the disc electrode the condition

$$\rho < 0{\cdot}1$$

must hold. This is equivalent to

$$\frac{\mathrm{d}i}{\mathrm{d}\eta} < 0{\cdot}36\, r_1 \kappa_\infty,$$

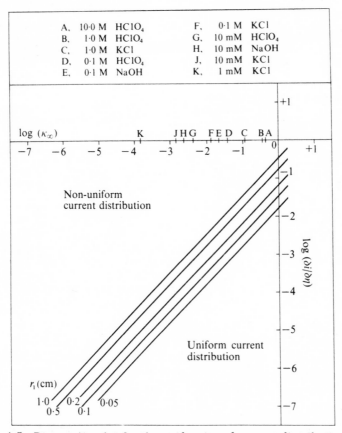

FIG. 4.5. Diagnostic plot for the uniformity of current distribution at an R.D.E. Some typical background electrolytes at 25°C in aqueous solution marked on log (κ_∞) scale in the figure. *Note:* $\mathrm{d}i/\mathrm{d}\eta$ is in units of Ω^{-1}; κ_∞ is in units of $\Omega^{-1}\mathrm{cm}^{-1}$.

Fig. 4.5 shows the critical values of $di/d\eta$ as a function of κ_∞ for different values of r_1. If $di/d\eta$ is smaller than the critical value the non-uniformity in current distribution will be negligible. It might be thought that a disc with a large radius would be preferable, but i is a total current and is therefore proportional to πr_1^2. Hence the reverse is true.

Experimental Tests of Non-Uniform Current Distribution

Experimental verification of these effects has been obtained in two ways. The first method is by depositing copper from a copper sulphate solution with only small concentrations of background

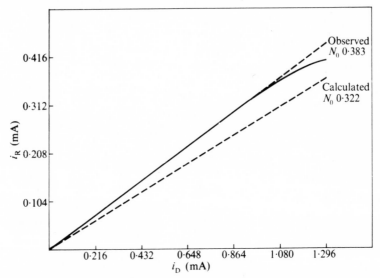

FIG. 4.6. Experimental $i_R - i_D$ trace for an R.R.D.E. under conditions of non-uniform current distribution at the disc electrode. [NaBr] = 3 mM; [HClO$_4$] = 1 mM. W = 20 Hz. Electrode geometry: r_1 = 0·3449 cm; r_2 = 0·3983 cm; r_3 = 0·4825 cm. *Note*: Curvature in experimental trace occurs because of oxidation of the Pt disc electrode at high disc currents.

electrolyte. This has been done by Marathe and Newman[3] and by Bruckenstein and Miller.[4] The thickness of the copper deposited was measured as a function of the radius of the disc and was found to be greater on the outside of the disc than at the centre. Reasonably good agreement was found by both sets of workers between theory and experiment.

The second method suggested by Albery and Ulstrup[5] involved the ring-disc electrode. The current at a thin-gap thin-ring electrode is determined mainly by the current density on the outside of the disc electrode. If this current density is larger than the average current density then values of N_o greater than that calculated from theory will be observed. Albery and Ulstrup did observe deviations but they were in the opposite direction to that predicted by Newman's theory. N_o was smaller than the calculated value. The reason for this has now been traced to the fact that the disc current was not 100 per cent efficient at producing the intermediate (Br_2); some oxidation of the platinum electrode was occurring at the same time. Fig. 4.6 shows more recent experimental results[6] in which, in qualitative agreement with theory, values of i_R/i_D greater than N_o are found. Similar results have been found by Bruckenstein and Miller.[4] Thus, experimentally, Newman's calculations have been confirmed. The main conclusion of this chapter must be to use Fig. 4.5 and keep out of trouble.

References

1. NEWMAN, J. (1966) *J. electrochem. Soc.* 113, 501.

2. NEWMAN, J. (1966) *J. electrochem. Soc.* 113, 1235.

3. MARATHE, V. and NEWMAN, J. (1969) *J. electrochem. Soc.* 116, 1704.

4. BRUCKENSTEIN, S. and MILLER, B. (1970) *J. electrochem. Soc.* 117, 1044.

5. ALBERY, W.J. and ULSTRUP, J. (1968) *Electrochim. Acta* 13, 281.

6. ALBERY, W.J. and HITCHMAN, M.L. (1969) unpublished results.

THE SCHEME OF SQUARES

Introduction

BEFORE discussing the application of the ring-disc electrode to electrode kinetics we will consider the fundamental problem of defining the mechanism of a complex electrode reaction. In Chapter 1 we considered a simple one-step electron transfer. This type of reaction is found in the classical inorganic redox systems. In general, however, most electrode reactions, especially those involving organic compounds, are not simple one-electron transfers. First, more than one electron may be transferred, and secondly, chemical steps may take place as well as electron transfers. We deal first with a simplified theory of what happens when two electrons are transferred, and secondly with the 'scheme of squares', which is useful for analysing the pathway through a network of electron transfers and chemical steps.

Two-Electron Reactions

We first write the general reaction

$$A \underset{k''_{-1}}{\overset{k'_1}{\rightleftharpoons}} [B] \underset{k''_{-2}}{\overset{k'_2}{\rightleftharpoons}} C.$$

Particular examples of this type of reaction [1,2] are

$$Tl(III) \overset{e}{\rightleftharpoons} [Tl(II)] \overset{e}{\rightleftharpoons} Tl(I)$$

and

$$Cu(II) \overset{e}{\rightleftharpoons} [Cu(I)] \overset{e}{\rightleftharpoons} Cu(0).$$

B, the intermediate, has been put into square brackets to show that $\Delta G°$ for the reaction

$$2B \longrightarrow A + C$$

is negative. If this is not so then B will be stable with respect to A and C and the overall electrode reaction will not be a simple two-electron transfer. There will be some marginal cases where $\Delta G° \sim 0$ which will lead to rather complicated behaviour, but this is outside the scope of this book.

We now make a second simplifying assumption, and that is that α, the transfer coefficient for each electron transfer, is $\frac{1}{2}$, i.e.

$$k'_{\pm n} = k'_{\pm n, E=0} \exp\left(\frac{\mp EF}{2RT}\right),$$

or

$$\log k'_{\pm n} = \log(k'_{\pm n, P=0}) \mp P, \tag{5.1}$$

where

$$P = EF/4\cdot606\,RT \tag{5.2}$$

and is a non-dimensional potential variable.

Fig. 5.1 shows a plot of the different forms for the variation of $\log k'_{\pm n}$ with P. The condition that $\Delta G° < 0$ for the reaction $2B \longrightarrow A + C$ means that

$$\log k'_2 - \log k'_{-2} > \log k'_1 - \log k'_{-1},$$

and this in turn means that in the rectangles of Fig. 5.1 the k'_2, k'_{-2} corner must be at more positive values of P (on the left) than the k'_1, k'_{-1} corner. This restricts the rectangles to the four drawn. The corner X in each rectangle is the point where $k'_2 = k'_{-1}$ and marks a shift in the position of the transition state. We have indicated this by using the symbols E1 and E2 to denote the potential regions where, respectively, the k'_1, k'_{-1} and the k'_2, k'_{-2} transition states are rate determining. For potentials more reducing than X the transition state is E1 while for potentials more oxidizing than X it is E2; note that X is not the same potential as Y. If k'_+ is the observed rate constant for A $\rightarrow$ C and k'_- that for C to A then

$$k'_2 \gg k'_{-1} \qquad\qquad k'_{-1} \gg k'_2$$

$$k'_+ = \frac{k'_1}{1 + k'_{-1}/k'_2}; \qquad \simeq k'_1 \qquad ; \qquad \simeq \frac{k'_1 k'_2}{k'_{-1}};$$

$$k'_- = \frac{k'_{-2}}{1 + k'_2/k'_{-1}}; \qquad \simeq \frac{k'_{-2}k'_{-1}}{k'_2} \qquad ; \qquad \simeq k'_{-2}.$$

In the diagrams we have plotted as full lines the observed k'_+ and k'_-. The slopes are ± 1 when the first electron transfer is rate determining, and ± 3 when there is a pre-equilibrium followed by a second rate-determining electron transfer; 3:1 is the ratio of $(1 + \alpha):\alpha$. For simplicity we have not plotted the full expression for k'_+ and k'_-, but

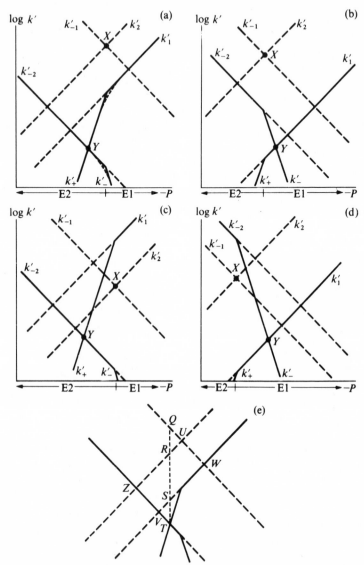

FIG. 5.1. Variation of individual rate constants and observed rate constant with potential for electrochemical reactions involving two electrons; dashed line for individual rate constants, solid lines for observed rate constants.

have used the two approximations. The use of the full expression would lead to $k'_{\pm}$ following a curve inside the obtuse angle as shown in Fig. 5.1(a).

The value of P at the point Y corresponds to the standard electrode potential. Here $k''_+ = k'_-$, and for equal activities of A and C no current would flow. The potential of Y is half way between the potentials of the k'_1, k'_{-1} and k'_2, k'_{-2} corners. This is because at the standard electrode potential

$$\log k''_1 - \log k'_{-1} = \log k'_{-2} - \log k'_2.$$

This is shown in a geometrical construction in Fig. 5.1(e). $QR = ST$, since QS ($\log k'_{-1} - \log k'_1$) represents the pre-equilibrium constant, and RT the effect of the pre-equilibrium on k'_2 to give the observed rate constant. From the parallel lines, $\triangle$s QRU and STV are congruent and hence the line $QRST$ is half way between U and V, and between W and Z. By inspection of Fig. 5.1(e) it can be seen that

$$\log k'_+ - \log k'_- = 4(P - P_o),$$

where P_o corresponds to E_o.

Ignoring activity coefficients then the condition for zero current is

$$k'_+ [A] = k'_- [C]$$

or

$$\log \frac{[A]}{[C]} = 4(P - P_o).$$

This is the Nernst equation for this system.

Relative Rates of Proton and Electron Transfer

We now consider the complications introduced by chemical steps such as proton transfer. For instance, if we consider the reduction of a quinone then the complete scheme of electron and proton transfers may be written[3,4] where A is a quinone

BH is a semiquinone

and CH_2 is a hydroquinone

The analysis into a scheme of squares has been discussed independently by Jacq[3] and by the authors.[4] In it the proton and electron transfers are assumed to take place separately. Many reductions of organic compounds probably take place in this fashion; for instance, the reduction of quinone to hydroquinone, or nitrobenzene to nitrosobenzene, to phenyl-hydroxylamine, to aniline. The chemical steps need not be proton transfers but might, for instance, for inorganic systems be the successive replacement of ligands. However, in the analysis that follows, we will restrict ourselves to proton transfers.

    One reason for this is that proton transfers between oxygen and nitrogen bases in aqueous solution are very rapid, and this simplifies the mathematics. If we consider the reduction of an acid-base pair HA and A in a solution buffered with X and HX then, ignoring the

convective term close to the electrode, the transport and kinetic
equations are

$$D\frac{\partial^2[A]}{\partial z^2} - k_1[A] + k_{-1}[HA] = 0$$

and

$$D\frac{\partial^2[HA]}{\partial z^2} + k_1[A] - k_{-1}[HA] = 0,$$

where

$$k_1 = k_{H_2O}^A + k_{H_3O^+}[H_2O^+] + k_{HX}[HX] \tag{5.3}$$

and

$$k_{-1} = k_{H_2O}^B + k_{OH^-}[OH^-] + k_X[X]. \tag{5.4}$$

We can solve these equations[5] to obtain a transport equation that gives

$$\frac{\partial([HA]+[A])}{\partial z} = \frac{([HA]+[A])_\infty - ([HA]+[A])_0}{Z_D} \tag{5.5}$$

and an equation describing the relaxation of the chemical equilibrium
with distance

$$k_1[A] - k_{-1}[HA] = (k_1[A]_0 - k_{-1}[HA]_0)\exp(-z/\mu), \tag{5.6}$$

where

$$\mu = \sqrt{\frac{D}{k_1 + k_{-1}}}$$

and describes the thickness of the reaction layer. Differentiating eqn
(5.6) at $z = 0$ we obtain

$$k_1\left(\frac{\partial[A]}{\partial z}\right)_0 - k_{-1}\left(\frac{\partial[HA]}{\partial z}\right)_0 = -\frac{k_1[A]_0 - k_{-1}[HA]_0}{\mu}. \tag{5.7}$$

On the electrode for an irreversible reaction

$$D\left(\frac{\partial[A]}{\partial z}\right)_0 = k_A'[A]_0 \tag{5.8}$$

and

$$D\left(\frac{\partial[HA]}{\partial z}\right)_0 = k_{HA}'[HA]_0. \tag{5.9}$$

Using eqns (5.7)–(5.9) we obtain

$$k_1\left(1 + \frac{k_A'\mu}{D}\right)[A]_0 = k_{-1}\left(1 + \frac{k_{HA}'\mu}{D}\right)[HA]_0. \tag{5.10}$$

If

$$\frac{k'_A \mu}{D} \ll 1 \quad \text{and} \quad \frac{k'_{HA} \mu}{D} \ll 1$$

then this equation reduces to the simple form

$$k_1 [A]_0 \simeq k_{-1} [HA]_0 ;$$

that is, the protonic equilibrium is labile enough so that the electron transfer steps do not alter the proportions of A and HA.

Now

$$\frac{k' \mu}{D} = \frac{k'}{\sqrt{D(k_1 + k_{-1})}}$$

and, in general,

$$k' < \sim 10 \text{ cm s}^{-1},$$

$$D \sim 10^{-5} \text{cm}^2 \text{s}^{-1}.$$

Thus

$$\frac{k' \mu}{D} < 1,$$

if

$$k_1 + k_{-1} > 10^7 \text{ s}^{-1}.$$

From eqns (5.3) and (5.4)

$$k_1 + k_{-1} = k^A_{H_2O} + k^B_{H_2O} + k_{H_3O^+} [H_3O^+] + k_{OH^-} [OH^-] +$$

$$k_{HX} [HX] + k_X [X].$$

Between a pH of 3 and 11 each of the terms $k_{H_3O} + [H_3O^+]$ and $k_{OH^-} [OH^-] < 10^7 \text{ s}^{-1}$, since the rate constants cannot exceed the diffusion controlled limit. Similarly if the pK of HA lies between 3 and 11 the rate constants involving H_2O will also be smaller than 10^7s^{-1}. Thus the lability of the proton transfer equilibrium in general depends upon the transfers to and from the buffer species HX and X.

Proton transfers between oxygen and nitrogen bases are generally diffusion controlled if the overall free energy for the transfer is favourable;[6] the rate constants are usually in the range 10^9 to $10^{11} M^{-1} s^{-1}$. Even when the overall free energy is zero, as for a symmetrical proton transfer, rate constants in this range are observed. Some values measured by N.M.R. are collected in Table (5.1). Hence if the [HX] and [X] ~ 0.1 M it is likely that in most systems the proton transfer steps will remain in equilibrium, and that the transition states will be the electron transfers. This conclusion can always be tested experimentally by keeping the buffer ratio constant but varying the total buffer concentration (see below).

TABLE 5.1

Rate constants for symmetrical proton transfers

A	HA	$k \times 10^{-9} \, (M^{-1} s^{-1})$	Ref.
H_2O	H_3O^+	10	7
NH_3	NH_4^+	1·3	8
$(CH_3)NH_3$	$CH_3NH_3^+$	0·9	8
OH^-	H_2O	5	7

The General Scheme of Squares

We will now discuss the form of the solution to the general problem of the n-unit square:

There are $2n^2$ unknowns; that is n^2 surface concentrations, $c_{p,q}$ and n^2 normalized concentration gradients on the electrode $g_{p,q}$ where

$$Z_D \left(\frac{\partial c_{p,q}}{\partial z} \right)_{z=0} = g_{p,q}.$$

To find these $2n^2$ unknowns we have, first, n^2 equations maintaining the steady state of each surface concentration. Note that this is a genuine steady state and we are not making the type of approximation that is made in homogeneous kinetics.

At the surface each species can undergo reduction or oxidation, or can diffuse back into the solution

$$\overset{k'_{q-1}}{\underset{k'_{-(q-1)}}{\rightleftharpoons}} \ \ A_{p,q} \ \overset{k'_q}{\underset{k'_{-q}}{\rightleftharpoons}} .$$

The steady-state conditions give

$$g_{p,q} \ = \ \lambda_{q-1} c_{p,q-1} - \lambda_{-(q-1)} c_{p,q} - \lambda_q c_{p,q} + \lambda_{-q} c_{p,q+1} ,$$

where

$$\lambda_q \ = \ k'_q Z_{\mathrm{D}}/D$$

and

$$\lambda_{\pm 0} \ = \ 0 \quad \text{and} \quad \lambda_{\pm n} \ = \ 0.$$

Secondly, we have n equations describing the transport of each group of species in chemical equilibrium through the diffusion layer to the electrode; that is, for each vertical column of species $A_{l,q}$

$$\sum_{l=1}^{l=m} g_{l,q} \ = \ \left(\sum_{l=1}^{l=m} c_{l,q} \right)_{\infty} - \left(\sum_{l=1}^{l=m} c_{l,q} \right)_0 .$$

Finally, we have $n(n-1)$ equations describing the chemical equilibria:

$$A_{p,q}$$
$$k_p \ \big\Updownarrow \ k_{-p}$$
$$A_{p+1,q} .$$

If the equilibria are all rapid enough for the condition discussed above to be fulfilled then there is one simple equation for each equilibrium,

$$c_{p+1,q} \ = \ K_p c_{p,q} .$$

Appendix 2 discusses the case when there are $(n-2)$ labile equilibria and one sluggish one. If more than one chemical step is sluggish then a more complicated situation prevails in which each equilibrium has its reaction layer thickness, μ_i. In general, the equilibria do not relax independently of each other. The differential equation can be solved but the algebra is more complicated.

We have not considered yet the case where μ and Z_{D} are comparable, or where $\mu > Z_{\mathrm{D}}$. Jacq has dealt with this case in more detail and has shown how it may be described using hyperbolic coth functions.[3] In Appendix 2 we have also dealt fully with the case of two linked equilibria.

Thus, to summarize, we have n^2 steady-state equations, n transfer equations, and $n(n-1)$ equilibrium equations with which to find $2n^2$ unknowns.

Even with $n = 2$ this gives eight simultaneous equations to be solved. We are not going to write them all out since, thanks to I.B.M., this is six more than we expect most individuals are able to solve today. Rather, we have tried to use a notation that will facilitate programming. Although the computer is better for numerical calculation, in the remainder of this chapter we present some three-dimensional blocks that may help us to understand the selection of particular routes through the scheme of squares.

One Square

$$
\begin{array}{ccc}
A & \xrightarrow{\;\;e\;\;} & B \\
& \overset{k'_1}{\underset{k'_{-1}}{\rightleftarrows}} & \\
k_1 \Big\Updownarrow k_{-1} & & k_2 \Big\Updownarrow k_{-2} \\
AH & \xrightarrow{\;\;e\;\;} & BH \\
& \overset{k'_2}{\underset{k'_{-2}}{\rightleftarrows}} &
\end{array}
$$

We consider the case where A and BH are the more stable species and AH and B are higher energy intermediates.

Each route can either have rate-determining electron or proton transfer. This depends upon whether the intermediate, either B or AH, is more likely to form A or BH. One cannot compare directly k'_{-1}, for instance, with k_2 because one is a heterogeneous and the other a homogeneous rate constant. Following the argument for eqn (5.10) the critical condition for the EC mechanism, $A \rightarrow B \rightarrow BH$, concerns the relative sizes of k'_{-1} and D/μ. For

$$k'_{-1} \gg \frac{D}{\mu}, \tag{5.11}$$

where

$$\mu = \sqrt{\frac{D}{k_2 + k_{-2}}},$$

we have pre-equilibrium electron transfer, and the proton transfer is rate determining:

$$\text{rate} = \frac{k'_1}{k'_{-1}} \frac{D}{\mu} [A]_0 = \frac{k'_1}{k'_{-1}} \{D(k_2 + k_{-2})\}^{1/2} [A]_0. \tag{5.12}$$

This has assumed that the proton transfers take place homogeneously in the solution, albeit close to the electrode. If the proton transfers take place directly from acid species in the solution on to species on the electrode, then instead of k_2, the homogeneous rate constant, we would have k_2'', a heterogeneous proton transfer rate constant (units cm s^{-1}). The condition would simply be

$$k_{-1}' \gg k_2'',$$

and the rate would be

$$\text{rate} = \frac{k_1'}{k_{-1}'} k_2'' [A]_0. \tag{5.13}$$

We have written first-order rate constants for the proton transfer reactions. In fact this will be composed of many terms:

$$k_2 + k_{-2} = k_{H_3O^+}[H_3O^+] + k_{HX}[HX] + k_{OH^-}[OH^-] + k_X[X] + k_{H_2O}.$$

As discussed above, it is likely that the terms involving the buffer will be the most important. If μ is determined by the downhill process

$$B + HX \xrightarrow{k_{HX}} BH + X,$$

then substitution for μ in eqn (5.12) gives

$$\text{rate} = \frac{k_1'}{k_{-1}'} D^{1/2} k_{HX}^{1/2} [HX]^{1/2} [A]_0. \tag{5.14}$$

By comparison, eqn (5.13) for heterogeneous proton transfer would give

$$\text{rate} = \frac{k_1'}{k_{-1}'} k_{HX,2}''' [HX][A]_0, \tag{5.15}$$

where $k_{HX,2}'''$ is a second-order heterogeneous rate constant (units cm^4 mol^{-1} s^{-1}). It is interesting that eqn (5.14) is half-order in [HX] while eqn (5.15) is simply first-order. Thus by varying the buffer components it may be possible to tell in the case of rate-determining proton transfer whether the transfer is taking place directly to species on the electrode or to species in the solution close to the electrode. The difference between the mechanisms is further discussed below.

Taking the opposite of condition (5.11) we have

$$k_{-1}' \ll \frac{D}{\mu}.$$

The rate-determining step is then the electron transfer and

$$\text{rate} = k_1' [A]_0.$$

For the CE route, $A \to AH \to BH$, the critical condition is determined by the relative sizes of k_2' and D/μ.

For $k_2' \ll D/\mu$

$$\text{rate} = K_1 k_2' [H^+][A]_0 . \tag{5.16}$$

In this case there is pre-equilibrium proton transfer followed by rate-determining electron transfer. The mechanism of the proton transfer is not important in this case.

On the other hand, for

$$k_2' \gg \frac{D}{\mu} ,$$

$$\text{rate} = K_1 \frac{D}{\mu} [H^+][A]_0 . \tag{5.17}$$

The rate-determining step is now the proton transfer. The rate is independent of the potential but does depend upon μ, the reaction layer thickness which in its turn, as discussed above, probably depends upon the concentration of the buffer components. In particular, the dominating term is likely to arise from the downhill process

$$AH + X \xrightarrow{k_X} A + HX$$

and so

$$\mu \simeq \sqrt{\frac{D}{k_X[X]}} .$$

Hence

$$\text{rate} = K_1 D^{1/2} k_X^{1/2} [X]^{1/2} [H^+][A]_0$$

$$= K_1 D^{1/2} k_X^{1/2} K_{HX}^{1/2} [H^+]^{1/2} [HX]^{1/2} [A]_0 . \tag{5.18}$$

We summarize the various mechanisms and the equations derived so far for the $A \to BH$ process, and we also write down, by similar arguments, the equations for the reverse process $BH \to A$, in Tables (5.2)–(5.4).

Heterogeneous vs. Homogeneous Proton Transfer

The difference between the heterogeneous and homogeneous proton transfers depends first upon the acidity or basicity of the species A, BH, etc. and secondly upon their adsorption on the electrode. If we make the assumption that the downhill proton transfers both for the heterogeneous and homogeneous processes are diffusion controlled and if we write ΔG_{ads} for the free energy of adsorption

$$A_{aq} \longrightarrow A_{electrode} ,$$

TABLE 5.2

Mechanistic criteria for the EC process, $A \to BH$ *and the CE process,* $BH \to A$

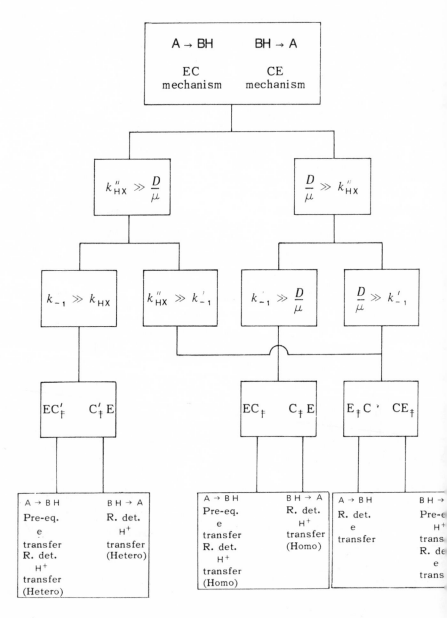

Note: Pre-eq. ≡ pre-equilibrium; R. det. ≡ rate determining.

TABLE 5.3

Mechanistic criteria for the CE process, $A \to BH$ *and the EC process,* $BH \to A$

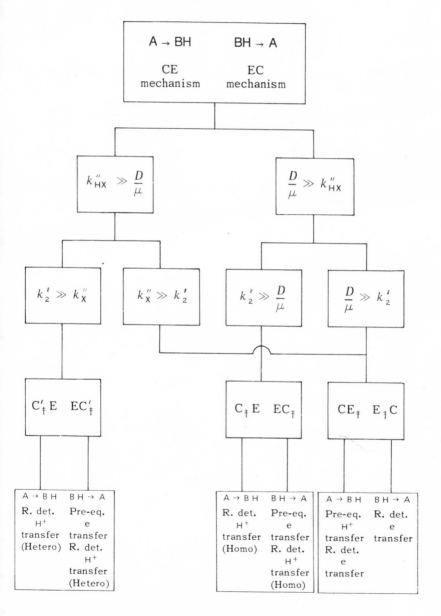

Note: Pre-eq. ≡ pre-equilibrium; R. det. ≡ rate determining.

c

TABLE 5.4

'Rate constants' for the processes A → BH *and* BH → A

	A → BH	BH → A
EC$_\ddagger'$	$\dfrac{k_1'}{k_{-1}'} k_{HX,2}'' \simeq \dfrac{k_1'}{k_{-1}'} k_{HX,2}''' [HX]$	$k_{X,2}'' \simeq k_{X,2}''' [X] = \dfrac{k_{HX,2}''' [HX]}{[H^+]}$
EC$_\ddagger$	$\dfrac{k_1'}{k_{-1}'} \dfrac{D}{\mu} \simeq \dfrac{k_1'}{k_{-1}'} D^{1/2} k_{HX}^{1/2} [HX]^{1/2}$	$\dfrac{D}{\mu K_2 [H^+]} \simeq \dfrac{D^{1/2} k_{HX}^{1/2} [HX]^{1/2}}{K_2 [H^+]}$
E$_\ddagger$C	k_1'	$\dfrac{k_1' K_2}{[H^+]}$
C$_\ddagger'$E	$k_{HX,1}'' \simeq k_{HX,1}''' [HX] = k_{X,1}''' [X][H^+]$	$\dfrac{k_{-2}'}{k_2'} k_{X,1}'' \simeq \dfrac{k_{-2}'}{k_2'} k_{X,1}''' [X]$
C$_\ddagger$E	$\dfrac{D}{\mu} K_1 [H^+] \simeq D^{1/2} K_1 k_X^{1/2} [X]^{1/2} [H^+]$	$\dfrac{k_{-2}'}{k_2'} \dfrac{D}{\mu} \simeq \dfrac{k_{-2}'}{k_2'} D^{1/2} k_X^{1/2} [X]^{1/2}$
CE$_\ddagger$	$k_2' K_1 [H^+]$	k_{-2}'

Notes: (1) The mechanistic labels are written for the process A → BH. The $\ddagger$ represents the rate determining step. C$'$ represents a heterogeneous proton transfer.

(2) $k'''_{,n}$ is a second-order heterogeneous rate constant for the nth equilibrium.

(3) $k_{HX,n}''' / k_{X,n}''' = K_n [H^+]$.

(4) The rate constants and equilibrium constants and the reactions to which they refer are summarized below.

Electron transfer

$$A \underset{k_{-1}'}{\overset{k_1'}{\rightleftharpoons}} B$$

$$AH \underset{k_{-2}'}{\overset{k_2'}{\rightleftharpoons}} BH$$

TABLE 5.4 (cont.)

Homogeneous proton transfers

$$AH + X \xrightarrow{k_X} HX + A$$

$$B + HX \xrightarrow{k_{HX}} X + BH$$

$$H + A \rightleftharpoons AH \qquad K_1$$

$$H + B \rightleftharpoons BH \qquad K_2$$

Heterogeneous proton transfers

$$A + HX \overset{k''_{HX,1}}{\underset{k''_{X,1}}{\rightleftharpoons}} X + AH$$

$$B + HX \overset{k''_{HX,2}}{\underset{k''_{X,2}}{\rightleftharpoons}} X + BH$$

then

$$k''_{HX} \text{ or } k''_X = \frac{D}{\mu},$$

when

$$c \sim \frac{10^3 \exp(2\Delta G_{ads}/RT)}{\bar{r}^3 L}$$

$$\sim 10^2 \exp(2\Delta G_{ads}/RT), \tag{5.19}$$

where c is the concentration of acid on base in mole litre^{-1}, $\bar{r}$ is the mean radius of A and the acid or base in centimetres, and L is Avogadro's number. This estimate can be derived by a number of arguments, one of which is given in Appendix 3. For $\Delta G_{ads} = 0$, the homogeneous process takes place for values of c below 10^2 M; this corresponds to one molecule per $\bar{r}^3 cm^3$. Thus the change-over takes place when, from any point on the electrode surface, there is a molecule of X or HX within the reaction distance. Below this concentration the substrate molecules diffuse into the solution before finding the buffer species; above it they get clobbered on the electrode. The critical concentration for $\Delta G_{ads} = 0$ can only be achieved by the solvent and we have the typical solvent value of 10^2 M. Hence strongly acidic or basic species which have diffusion controlled reactions with water will react by the heterogeneous route.

Species with $2 < pK < 12$ will tend to react by the homogeneous route unless ΔG_{ads} is rather negative. For example, a

$\Delta G_{ads} \sim -25\,\mathrm{kJ\ mol^{-1}}$ or $-6\,\mathrm{kcal\ mol^{-1}}$ would cause a mechanistic change at $c \sim 10^{-2}\mathrm{M}$.

Thus, to summarize,

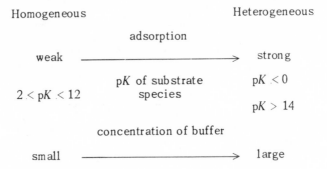

Shift in Mechanism with Potential

To discuss the shift in mechanism round the square with potential we now list the dependence (n) of the different rates on

$$P = \frac{1}{4 \cdot 606} \frac{EF}{RT}$$

in Table (5.5), which is obtained from Table (5.4).

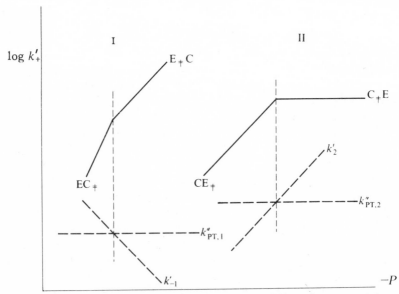

FIG. 5.2. Variation of rate constant for $A \rightarrow BH$ with potential for different routes round one square.

TABLE 5.5

Value of n for rate constants for one square

$$\log k' = \log k_0^{\,2} - nP$$

Mechanism	k'_+ $A \to BH$	k'_- $BH \to A$	Condition	
EC $EC^{\,i}_{\ddagger}$ ⎫ ⎬ $EC_{\ddagger}$ ⎭	2	0	$k'_{-1} \gg k''_{PT,1}$ $k'_{-1} = k''_{PT,1}$	n for k'_{-1} $= -1$
$E_{\ddagger} C$	1	-1	$k'_{-1} \ll k''_{PT,1}$	
CE $C^{\,i}_{\ddagger} E$ ⎫ ⎬ $C_{\ddagger} E$ ⎭	0	-2	$k'_2 \gg k''_{PT,2}$ $k'_2 = k''_{PT,2}$	n for k'_2 $= +1$
$CE_{\ddagger}$	1	-1	$k'_2 \ll k''_{PT,2}$	

Note: $k''_{PT,n}$ are proton transfer rate constants in cm s^{-1} (e.g. D/μ) and are independent of potential.

The dependence of k'_+ on P for the EC and CE mechanisms are drawn in Fig. 5.2 as lines I and II. The faster of these two routes will be the one observed at any potential. At large negative P this must be $E_{\ddagger} C$, and at large positive P, $CE_{\ddagger}$; that is, the electron transfer processes are always rate determining at extremes of potential.

In between, the mechanistic routes can intersect in a number of ways, shown in Fig. 5.3, for both the reduction $A \to BH$ and the oxidation $BH \to A$. For any particular system the $A \to BH$ curve will intersect the $BH \to A$ curve at the equilibrium potential for a solution containing equal activities of A and BH. A new system of labelling the rate-determining process has also been introduced in Fig. 5.3:

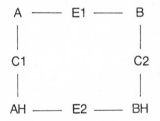

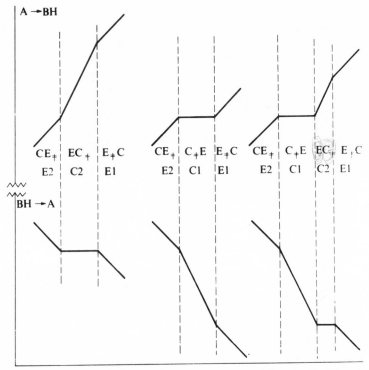

FIG. 5.3. Variation of observed rate constants with potential for
different possible combinations of routes round one square.

It is interesting to note that since the mechanism shifts with electrode
potential it is possible for the process $A \to BH$ to take place at reducing
potentials round the EC side of the square, while the reverse process
$BH \to A$ comes back at oxidizing potentials round the other side of the
square. In general, it is therefore dangerous to extrapolate mechanisms
out of the potential range in which they have been studied. One further
point worth noting is that the change from an EC to a CE mechanism
cannot take place directly from E2 to E1, but involves a potential
range where a chemical step is rate determining.

Shift in Mechanism with pH

As we have discussed above (Table 5.4), there are many possi-
bilities for the exact form of the equations governing proton transfer.
However, we will take the simplest case for our picture of the different
routes round the square. We will assume that experiments are carried
out in solutions where the $[HX]$ and the $[X]$ are constant, and that the
buffer components control either μ for homogeneous transfer or k'' for

heterogeneous transfer; we also assume that all the downhill rate
constants are diffusion controlled and are equal; i.e. $k_{HX} = k_{HY}$ etc.,
or $k''_{HX,2} = k''_{HY,2}$ etc. These assumptions are equivalent to assuming
that μ is a constant. They are also convenient in that the pH dependence
of D/μ and k'' becomes the same so that the same picture will serve
for both homogeneous and heterogeneous proton transfers; the breaks
between E2 and C1 and C2 and E1 in Fig. 5.3 will not depend on pH.
Although we are treating the simplest case here, the principles of the
analysis hold for more complicated cases, for example, unbuffered
solutions of H_3O^+.

As before, we start by listing the order of the different rate
constants with respect to H^+ using Tables 5.4–5.6. The effect of

TABLE 5.6

Order of rate constants with respect to H^+

Mechanism	$A \rightarrow BH$	$BH \rightarrow A$
EC		
C2 ($EC_\ddagger$)	0	-1
E1 ($E_\ddagger C$)	0	-1
CE		
C1 ($C_\ddagger E$)	$+1$	0
E2 ($CE_\ddagger$)	$+1$	0

increasing pH is to shift the CE part of the curve downwards with
respect to the EC part, as shown in Fig. 5.4. As this happens, two
patterns of behaviour can be revealed, as shown in the strip cartoons
in Fig. 5.4. Fig. 5.5 shows a three-dimensional block for each type of
behaviour, together with the ground plan that gives the change in rate-
determining process as a function of the potential and pH. The blocks
have the same fundamental shape for the reverse process $BH \rightarrow A$, except
that the axes are altered as shown. Which of the two different types
of behaviour (I or II, Fig. 5.4) is observed depends upon the following
condition: if at the potential where $k'_{-1} = k'_2 = k'_*$, k''_{PT} is greater than
k'_*, then scheme I will be found; but if at this potential $k''_{PT} < k'_*$ then
scheme II will be found. The reason for this is shown in Fig. 5.4. Since
proton transfers are rapid in buffered systems it is more likely that
$k''_{PT} \gg k'_*$ and therefore scheme I is more likely to be found than
scheme II. Fig. 5.6 shows how a block for the process $A \rightarrow BH$ inter-
sects with a block for $BH \rightarrow A$. The intersection gives the equilibrium
potential for solutions with equal activities of A and BH as a function
of pH. We have drawn the intersection passing first near the centre of

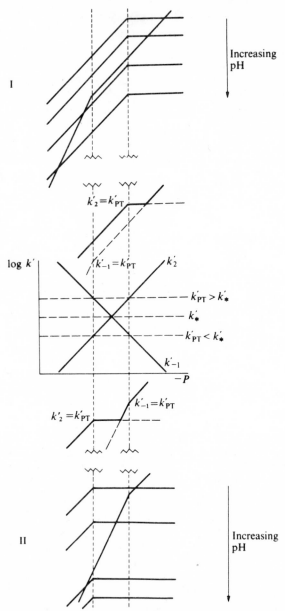

FIG. 5.4. Effect of pH on observed rate constant for $A \rightarrow BH$.

the ground plan and secondly when it is well off centre. It may be seen
that the more that the inequality $k_{PT}'' \gg k_*'$ obtains the faster will be
the proton transfer with respect to the electron transfers, the boundary

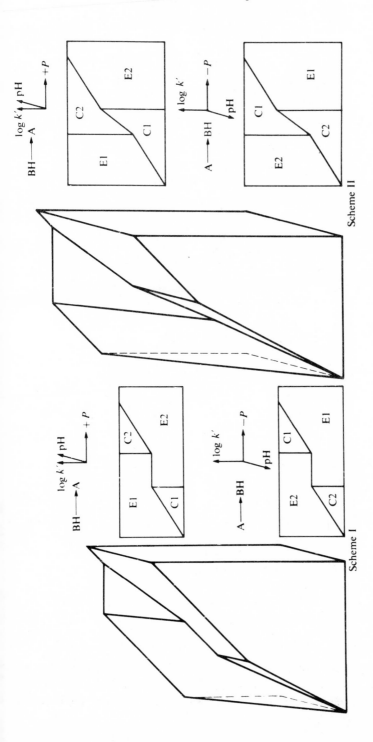

FIG. 5.5. Blocks showing effect of potential and pH on observed rate constant for A → BH and for BH → A for schemes I and II.

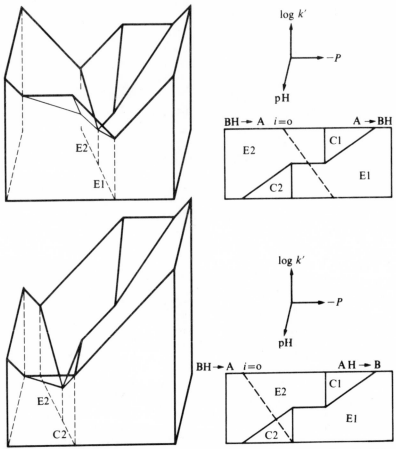

FIG. 5.6. Blocks showing effect of potential and pH on $A \rightleftharpoons BH$ system for scheme I.

between E1 and E2 in scheme I will be extended, and it may well be that all the observed behaviour of the system will be found in the valley formed by the E1 and E2 planes. This corresponds to the proton transfers being in rapid equilibrium and the different electron transfers being rate determining. This is shown in Fig. 5.7. The value of pH which marks the E1–E2 boundary is the value of the pK of the E2 transition state treated as an acid. It is not surprising that we have the protonated form of the transition state (E2) for pH $<$ pK_{E2} and the unprotonated form (E1) for pH $>$ pK_{E2}.

Returning to Fig. 5.6, in practice we cannot reach the remote C1 and C2 regions since at the large overpotentials necessary the currents would have become limited by transport.

This word has been neglected for a long time; however the effects of transport are easily included. Using a similar argument to eqn (2.13) we obtain

$$ j \ = \ \frac{k'_+[A]_\infty - k''_-[BH]_\infty}{1 + \dfrac{Z_D}{D}(k'_+ + k''_-)} \ . $$

Thus the determination of k'_+ and k'_- determines j.

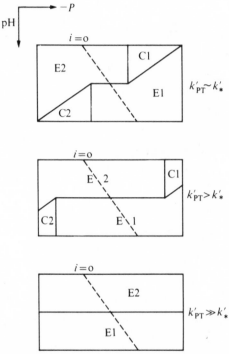

FIG. 5.7. The effect on scheme II of increasing the rates of proton transfer with respect to the electron transfer rates.

Three-unit Square

We now combine the two-electron transfer section with the sections on the one square to consider the system

A —— E1 —— B —— E2 —— C

AH —— E3 —— BH —— E4 —— CH

AH$_2$ —— E5 —— BH$_2$ —— E6 —— CH$_2$

In this system we have made the simplifying assumption of rapid proton transfers, and we follow that with some more assumptions about the pKs of the various species that are likely to be found in practice.

First,

$$pK_{AH_2} < pK_{AH}, \qquad pK_{E5} < pK_{E3};$$

$$pK_{BH_2} < pK_{BH}, \qquad pK_{E6} < pK_{E4};$$

$$pK_{CH_2} < pK_{CH};$$

and secondly,

$$pK_{AH} < pK_{E3} < pK_{BH} < pK_{E4} < pK_{CH}; \qquad (5.20)$$

$$pK_{AH_2} < pK_{E5} < pK_{BH_2} < pK_{E6} < pK_{CH_2}. \qquad (5.21)$$

This second assumption assumes that the acid strength of the species decreases steadily with increasing addition of electrons. The pK's of the transition states then fall into two possible orders:

$$pK_{E4} > pK_{E6} > pK_{E3} > pK_{E5} \qquad \text{III} \qquad (5.22)$$

and

$$pK_{E4} > pK_{E3} > pK_{E6} > pK_{E5}. \qquad \text{IV} \qquad (5.23)$$

The difference between the schemes depends upon

$$pK_{E6} \qquad (BC)H_{2\ddagger} \rightleftharpoons (BC)H_{\ddagger} + H^+$$

and

$$pK_{E3} \qquad (AB)H_{\ddagger} \rightleftharpoons AB_{\ddagger} + H^+,$$

or on ΔG for

$$AB_{\ddagger} + (BC)H_{2\ddagger} \rightleftharpoons (AB)H_{\ddagger} + BCH_{\ddagger}$$
$$\text{E1} \qquad \text{E6} \qquad \text{E3} \qquad \text{E4} \quad .$$

If ΔG is $+$ve then we have scheme III and if ΔG is $-$ve, scheme IV. As shown below, scheme III is associated with the CEEC mechanism through E3 and E4, while scheme IV prefers the ECCE mechanism through E1 and E6.

The possible routes round the square together with their transition states and pH conditions corresponding to schemes III and IV are given in Table (5.7).

The assumptions (5.20) and (5.21) ruled out tortuous routes of the type E3–E2 (CECECC), since it is impossible with the conditions (5.22) and (5.23) to satisfy the condition

$$pK_{E3} > pH > pK_{E4},$$

which is required by E3–E2. Thus the assumptions (5.20) and (5.21) lead to the simple, direct routes from A to CH_2.

TABLE 5.7
Routes round a 3×3 square

Mechanism	Transition states	pH condition	Scheme
EECC	E1, E2	$pH > pK_{E4}$	III and IV
ECEC	E1, E4	$pK_{E4} > pH > pK_{E6}$	III
		$pK_{E4} > pH > pK_{E3}$	IV
ECCE	E1, E6	$pK_{E6} > pH > pK_{E3}$	III
CEEC	E3, E4	$pK_{E2} > pH > pK_{E6}$	IV
CECE	E3, E6	$pK_{E3} > pH > pK_{E5}$	III
		$pK_{E6} > pH > pK_{E5}$	IV
CCEE	E5, E6	$pK_{E5} > pH$	III and IV

Each route is now reduced to a two-electron reaction similar to those discussed at the beginning of the chapter. As seen in Fig. 5.1 for reducing potentials, E1, E3, and E5 of each pair will be the rate-determining electron transfer, while for oxidizing potentials it will be E2, E4, and E6. Following the arguments used for the one square we can now construct the blocks and ground plans which show the variation of rate constant for the different mechanisms for scheme III and scheme IV as a function of pH and electrode potential. These are shown in Figs. 5.8 and 5.9. Each plane represents a different transition state and, depending on the pH's that divide E1 from E3 and E4 from E6, two different patterns are found. For convenience the critical pH's have been drawn equally spaced; this is unlikely to be found for real systems but the essential shapes and patterns still remain. The block represents the rate constant in one direction. A similarly shaped block but with the pH and potential axes reversed will be found for the rate constant for the reverse process. Thus any system consists of an intersection of two of these blocks, one for $A \rightarrow CH_2$ and one for $CH_2 \rightarrow A$.

Each block has a 'hinge' which divides the shallow planes $(\partial \log k'/\partial P = 1)$ on the right from the steep planes $(\partial \log k'/\partial P = 3)$ on the left. On the ground plans the projection of the hinge divides E1, E3, and E5 from E2, E4, and E6. This hinge marks the shift in the transition state from the first electron transfer being rate determining (shallow planes on the right) to a pre-equilibrium followed by a second rate-determining electron transfer (steep planes, on the left). It marks the same type of shift as discussed in Fig. 5.1. The variation of $\log k'$ with pH depends upon the number of chemical steps preceding the rate-determining electron transfer. The gradient $\partial \log k'/\partial$ pH is zero for the front pair of planes (no pre-equilibrium proton transfer), increases to one for the middle plane (one pre-equilibrium), and ends at two at the back of the block (two pre-equilibria). As can be seen from the

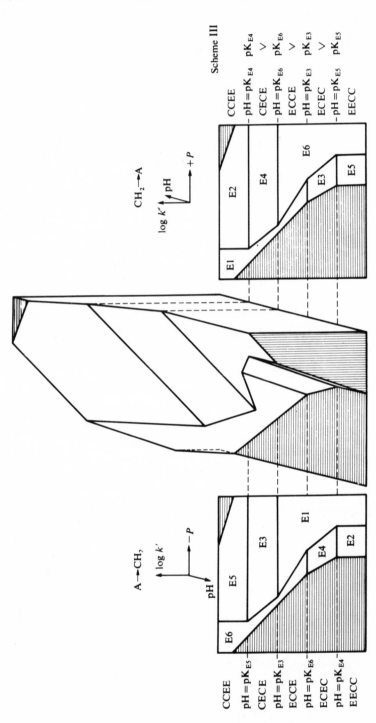

FIG. 5.8. Block showing effect of potential and pH on forward or back-ward rate constant for scheme III for A, CH_2 system.

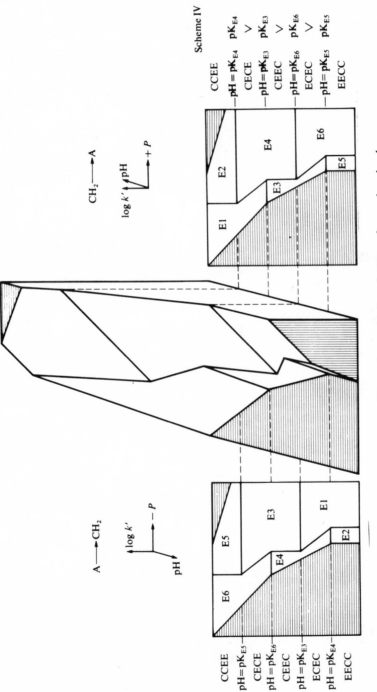

FIG. 5.9. Block showing effect of potential and pH on forward or backward rate constant for scheme IV for A, CH₂ system.

ground plans, the 'hinge' does not alter with pH for the pairs E1 and E2, E3 and E4, and E5 and E6, which have the same number of pre-equilibria. As one might expect, the reaction $A \rightarrow CH_2$ occurs more rapidly in acid solution while $CH_2 \rightarrow A$ goes faster at high pH.

Fig. 5.10 shows the intersection of two blocks for the scheme IV (CEEC) mechanism. A symmetrical intersection has been drawn. The

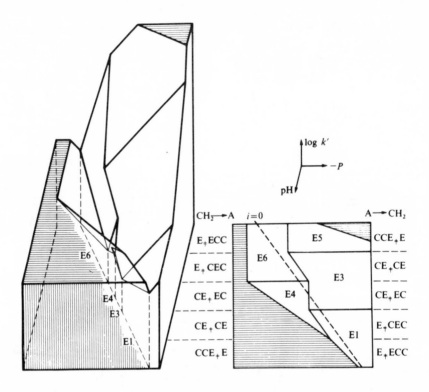

FIG. 5.10. Block showing effect of potential and pH on observed rate constants for $A \rightleftharpoons CH_2$ for scheme IV with a symmetrical intersection.

block on the right is the same as the block in Fig. 5.8. The one on the left has its top lopped off. The intersection of the blocks gives the equilibrium potential for unit activities of A and CH_2 as a function of pH. It obeys the Nernst equation. Along this intersection $\log k'$ falls with pH in the E6 region, rises to the pass in the E4 region, crosses into E3 and falls again, and finally rises towards the front of the block in the E1 region. The hinge that divides an E_{2n+1} transition state from an E_{2n} starts at the back on the right-hand side of the valley, crosses the intersection at the maximum in $\log k'$, and ends up on the left-hand side.

In Fig. 5.11 a more asymmetric intersection has been drawn. This has been done by keeping the left-hand block constant and raising the right-hand block by five $\log k'$ units. The intersection now no longer crosses as many of the right-hand planes but remains in E4 for a wider range of pH. (It still obeys the Nernst equation.) The value of $\log k'$ passes through a minimum on the E6, E4 border. As the right-hand block is raised still further to give a more asymmetric intersection

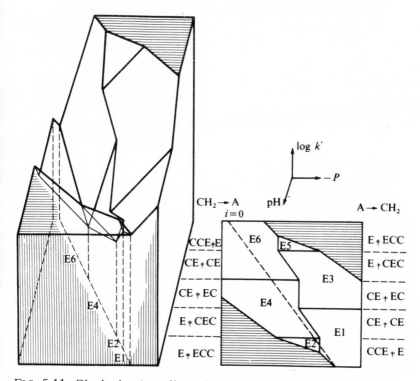

FIG. 5.11. Block showing effect of potential and pH on observed rate constants for $A \rightleftharpoons CH_2$ for scheme IV with an asymmetrical intersection.

the shallow planes (E_{2n+1}) move further and further away from the intersection. We have not drawn what happens when the right-hand block drops below the left-hand one; but take another look at Fig. 5.11 with the right-hand block as the $CH_2 \rightarrow A$ processes and the pH and P axes reversed.

Fig. 5.12 shows the symmetrical intersection of two blocks for scheme III (ECCE). Again the intersection obeys the Nernst equation.

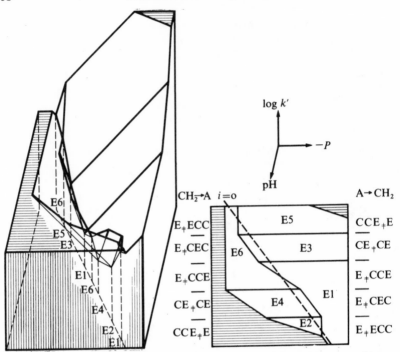

FIG. 5.12. Block showing effect of potential and pH on observed rate constants for A $\rightleftharpoons$ CH$_2$ for scheme III with a symmetrical intersection.

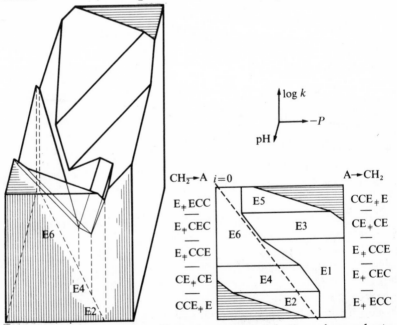

FIG. 5.13. Block showing effect of potential and pH on observed rate constants for A $\rightleftharpoons$ CH$_2$ for scheme III with an asymmetrical intersecti

The behaviour of $\log k'$ reminds us of the Küblis run. The various sections may be listed starting from the Weissfluhjoch:

E6	E5	E3	E1	E6	E4	E2	E1
down	steeply down	down	up	down	up	steeply up	up

There are seven changes of transition state as the pH changes. Fig. 5.13 shows the left-hand block the same but the right-hand one raised by four $\log k'$ units to give a more asymmetric intersection. The behaviour of the intersection is now much simpler, going through a minimum on the E4, E6 border. Its crossing into the E1 region takes place between the reader and the page.

These blocks have been constructed as plots of $\log k'$ against P, where the e.m.f. is measured against a reference electrode (e.g. the S.C.E.). They can equally well be interpreted as plots of $(\log k'[A]_o.)$ and $\log(k'[CH_2]_o.)$ against P' where

$$P' = \frac{\eta F}{4 \cdot 606\,RT};$$

$P' = 0$ on the intersection and η is the overpotential. In this interpretation changing the ratio of the surface concentrations by a factor of f shifts one block relative to the other by f units.

For those without developed stereoscopic vision Fig. 5.14 shows some unit blocks for each type of route through the scheme of squares.

Although these blocks show how one mechanism shifts into another it is unlikely that any real system will show all of this behaviour. There are restrictions on all axes. On the pH axis the dissociation constant of some of the transition states may be outside a convenient range for study; the use of strong acids or bases leads to problems of non-ideality and the onset of electrode reactions of H^+ and OH^-. On the e.m.f. axis we are restricted by the stability of the solvent and the other species in the solution to electrochemical decomposition. On the $\log k'$ axis the rate must be large enough to distinguish from residual currents, but not so large that the currents become transport rather than kinetically controlled. This means that typically for k' (in cm s^{-1}):

$$10^{-4} < k' < 10^{-1}.$$

Thus on the vertical axis we are restricted to a slice of the block. For a reversible system this must be close to the equilibrium potential, where the rate constants are low enough to come into the critical slice. For an irreversible system, on the other hand, the value of $\log k'$ in the valley may be too low to be measured accurately and we may be restricted to values high up on the sides of the pass.

We finally summarize some of the important features of this discussion.

(1) At extreme reducing and oxidizing potentials the first electron transfer will be rate determining.

(2) These two regions are separated by a region where there is a pre-equilibrium separated by a rate-determining second electron transfer.

(3) The shift of mechanism with pH depends on the pK's of the transition states.

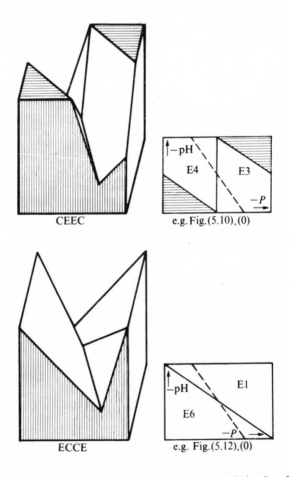

FIG. 5.14. Building blocks for different routes round the 3×3 square. In the examples the bracket after the figure number indicates when the axes have been changed.

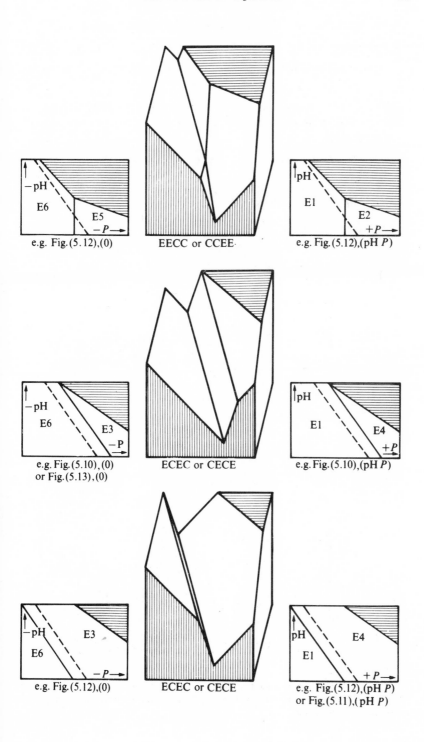

e.g. Fig.(5.12),(0) EECC or CCEE e.g. Fig.(5.12),(pH P)

e.g. Fig.(5.10),(0)
or Fig.(5.13),(0) ECEC or CECE e.g. Fig.(5.10),(pH P)

e.g. Fig.(5.12),(0) ECEC or CECE e.g. Fig.(5.12),(pH P)
or Fig.(5.11),(pH P)

(4) These shifts are likely to follow one of two patterns:

Scheme III			Scheme IV	
$A \rightarrow CH_2$	$CH_2 \rightarrow A$		$A \rightarrow CH_2$	$CH_2 \rightarrow A$
CCEE	EECC	pH	CCEE	EECC
CECE	ECEC	↑	CECE	ECEC
ECCE	ECCE		CEEC	CEEC
ECEC	CECE	↓	ECEC	CECE
EECC	CCEE	increasing	EECC	CCEE

This discussion has ignored a number of important features of electrode mechanisms; for instance, the role of the double layer, adsorption, and more complicated reaction orders. This is partly because these matters have been dealt with elsewhere,[9,10] and partly because the ring-disc electrode is particularly pertinent to the study of intermediates and the elucidation of multi-step mechanisms.

References

1. ULSTRUP, J. (1968) *Electrochim. Acta* 13, 535.

2. NEKRASOV, L.N. and BEREZINA, N.P. (1962) *Dokl. Akad. Nauk SSSR* 142, 855.

3. JACQ, J. (1967) *Electrochim. Acta* 12, 1345.

4. FIELD, N.J.R. (1968) *D. Phil. Thesis*, Oxford.

5. KOUTECKY, J. and LEVICH, V.G. (1958) *Zh. fiz. Khim.* 32, 1565.

6. EIGEN, M. (1963) *Angew. Chem.* 75, 489.

7. MEIBOOM, S. (1961) *J. chem. Phys.* 34, 375.

8. GRUNWALD, E. and COCIVERA, M. (1965) *Discuss. Faraday Soc.* 39, 105.

9. CONWAY, B.E. (1965) *Theory and principles of electrode processes* Ronald Press, New York.

10. DELAHAY, P. (1965) *Double layer and electrode kinetics.* Interscience.

6

ELECTRODE KINETICS

Introduction

FROM the previous chapter it can be seen that the determination of the mechanism of an electrode reaction may be rather complex. For instance, for the reduction of a quinone to a hydroquinone twelve different possible rate-determining processes have to be considered. Even if the route for the reduction is established the oxidation may return along a different path. The ring-disc electrode can provide valuable information about the nature of the intermediates formed in the electrode mechanism, their dependence on pH, and the rates of their formation and decomposition.

The Ivanov—Levich Equation

When the reduction A to CH_2 is taking place on the electrode the intermediates B, BH, and BH_2 may have a long enough lifetime for them to diffuse off the disc and be detected at the ring electrode. If we take, to start with, the simple system

$$\text{disc} \quad A \xrightarrow{n_1 e} B \xrightarrow{n_2 e} C$$

$$\text{ring electrode}$$

$$\text{ring} \quad B \xrightarrow{n_3 e} D,$$

we follow the original argument of Ivanov and Levich.[1] The disc current will be given by

$$i_D = AFD \left\{ (n_1 + n_2) \left(\frac{\partial a}{\partial z} \right)_o + n_2 \left(\frac{\partial b}{\partial z} \right)_o \right\}. \tag{6.1}$$

The ring current is found by using the argument of Chapter 3 with

$$u = \text{a constant} \times b,$$

where the constant is chosen so that $(\partial u / \partial w)_{\substack{w=0 \\ r < r_1}} = -1$. We then obtain

73

the general result

$$i_R = -N_o \, n_3 \, AFD \left(\frac{\partial b}{\partial z}\right)_{\substack{o \\ r < r_1}} \qquad (6.2)$$

where, since $b = 0$ from the transport equation for b for $r < r_1$,

$$\left(\frac{\partial b}{\partial z}\right)_o = -\frac{b_o}{Z_D}. \qquad (6.3)$$

On the disc electrode, since b is in a genuine steady state,

$$D\left(\frac{\partial a}{\partial z}\right)_o = -D\left(\frac{\partial b}{\partial z}\right)_o + k' \, b_o. \qquad (6.4)$$

Using eqns (6.1)–(6.4) to eliminate b_o, $(\partial a/\partial z)_o$, and $(\partial b/\partial z)_o$ we obtain

$$i_R = \frac{N_o \, n_3 \, i_D}{n_1 + (n_1 + n_2)\, \dfrac{k' \, Z_D}{D}}. \qquad (6.5)$$

When $k' \ll D/Z_D$ and $n_3 = -n_1$ we obtain the usual result:

$$i_R = -N_o \, i_D ,$$

since all the B escapes into solution; (the negative sign arises because one current is oxidizing and the other reducing).

On the other hand, when $k' \gg D/Z_D$ the B reacts further on the disc electrode and very little arrives at the ring electrode. Increasing the rotation speed lowers Z_D ($Z_D \propto \omega^{-1/2}$) and more B will escape into the solution.

For the scheme of squares discussed in the last chapter

$$
\begin{array}{ccc}
\text{A} \longrightarrow \text{B} \xrightarrow{k'_{1,2}} \text{C} \\
\Updownarrow \quad K_{1,2}\Updownarrow \qquad \Updownarrow \\
\text{AH} \longrightarrow \text{BH} \xrightarrow{k'_{2,2}} \text{CH} \\
\Updownarrow \quad K_{2,2}\Updownarrow \qquad \Updownarrow \\
\text{AH}_2 \longrightarrow \text{BH}_2 \xrightarrow{k'_{3,2}} \text{CH}_2
\end{array}
$$

for the complete solution we can calculate $\partial([\text{B}] + [\text{BH}] + [\text{BH}_2])/\partial z$ at $z = 0$ and use eqn (6.2) to relate this to the ring current. If we make the same assumptions as were made in the last chapter, that is that the proton transfers are rapid, then we obtain an equation of the same form as eqn (6.5) except that k' is made up of the individual processes and is pH-dependent:

$$k' = \frac{k'_{1,2} + K_{1,2}\,k'_{2,2}[H^+] + K_{1,2}\,K_{2,2}\,k'_{3,2}[H^+]^2}{1 + K_{1,2}\,[H^+] + K_{1,2}\,K_{2,2}\,[H^+]^2}.$$

It can be seen that the pH-dependence of a ring current may yield information not only about the rate processes but also about $K_{1,2}$ or $K_{2,2}$.

The ring-disc electrode can also be used to investigate branching in electrode mechanisms. Damjanovic, Genshaw, and Bockris[2] have developed a diagnostic test to tell whether branching is taking place or not. We have the general scheme

$$\text{disc} \quad A \xrightarrow{n_1} B \xrightarrow[k']{n_2} C$$

$$\downarrow$$

ring electrode

$$\text{together with} \quad A \xrightarrow{n_4} D.$$

$$\text{ring} \quad B \xrightarrow{n_3} E.$$

The test depends upon varying the rotation speed while keeping the potential constant. It is assumed that although the branching between the formation of B or D from A may be sensitive to potential, it is not affected by the rotation speed. We define the ratio x where

$$x = \frac{\text{flux of } A \longrightarrow D}{\text{flux of } A \longrightarrow B}. \tag{6.6}$$

The disc current is then given by

$$i_D = AFD\left\{ \frac{(n_1 + n_2 + xn_4)}{1 + x}\left(\frac{\partial a}{\partial z}\right)_o + n_2\left(\frac{\partial b}{\partial z}\right)_o \right\},$$

and this leads by the same argument as above to

$$\frac{N_o i_D}{i_R} = \frac{n_1 + xn_4}{n_3} + \frac{n_1 + n_2 + xn_4}{n_3}\frac{k'}{D}A_3 W^{-1/2}, \tag{6.7}$$

where $Z_D = A_3 W^{-1/2} = 0{\cdot}643\,\nu^{1/6}D^{1/3}W^{-1/2}$. A plot of the left-hand side against $W^{-1/2}$ should give a straight line, from which x and k' can be determined. Diagnostically we find the results given in Table (6.1). Care must be taken in using these tests since it is possible for other reaction schemes, not considered here, to give the same results. For example, a scheme of the type

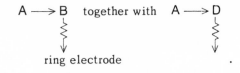

where B is electroactive on the ring and C is not, will give a similar plot to the scheme

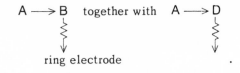

TABLE 6.1

Results for the diagnostic equation (6.7)

Reaction scheme on disc	Parameters	Gradient	Potential dependent	Intercept	Potential dependent
$A \longrightarrow B \longrightarrow C$	$k' \sim D/Z_D \,; x = 0 \; > 0$		Yes	$\dfrac{n_1}{n_3}$	No
$\left.\begin{array}{l} A \longrightarrow B \\ \\ A \longrightarrow D \end{array}\right\}$	$k' = 0 \,; x \sim 1$	0	No	$\dfrac{n_1 + xn_4}{n_3}$	Yes
$\left.\begin{array}{l} A \longrightarrow B \longrightarrow C \\ \\ A \longrightarrow D \end{array}\right\}$	$k' \sim D/Z_D \,; x \sim 1 \; > 0$	Yes		$\dfrac{n_1 + xn_4}{n_3}$	Yes

Split-ring Electrode

Miller and Visco[3] have developed a further refinement on the basic ring-disc electrode which is particularly useful for the study of electrode kinetics. They have divided the ring electrode into two semi-circular segments as in Fig. 6.1. The potential of each half of the ring electrode can be independently controlled. Thus the stream of material from the disc electrode can be simultaneously monitored at two different potentials. There are two advantages of simultaneous monitoring; first it saves time and secondly, and more important, experimental error from time-dependent behaviour on the disc electrode is eliminated. Solid electrodes are not well behaved but we can draw mechanistic

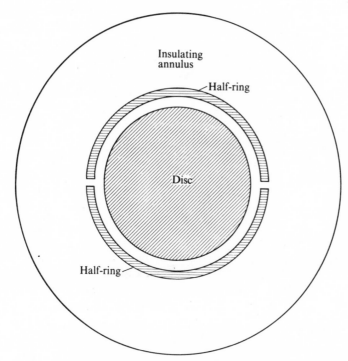

FIG. 6.1. Split ring-disc electrode.

conclusions if we can obtain sufficient information at any one time. Reasons for using two potentials may be that we wish to characterize the oxidation state of a single intermediate by simultaneous oxidation and reduction, or that we have two intermediates which have different half-wave potentials and so their rates of generation by the disc electrode can each be measured.

Since with respect to the disc the solution close to it is stagnant, there is little interference between one half-ring and the other. This can be tested by measuring a 'collection inefficiency':

one half-ring $A \longrightarrow B$,

other half-ring $B \longrightarrow A$.

Typically, values < 0.0005 are obtained. The collection efficiency of each ring with respect to the disc is about $\frac{1}{2}N$ but should be measured for each electrode. If the two half rings are yoked together at the same potential then the usual value of N_0 is obtained. Table 6.2 shows some data for gold split-ring electrode used in Oxford.[4]

TABLE 6.2

Collection efficiencies of a split ring-disc electrode

Electrode geometry $r_1 = 0.361$ cm; $r_2 = 0.388$ cm; $r_3 = 0.411$ cm.

Distance between two half-rings 0.0275 cm.

Experimental collection efficiencies:

       one half-ring $N_{\frac{1}{2}} = 0.0798$,

       other half-ring $N_{\frac{1}{2}} = 0.0869$,

       half-rings yoked $N_0 = 0.1667$.
         together

Theoretical collection efficiency $N_0 = 0.1669$.

The Copper System

One of the first systems studied by the ring-disc electrode was the reduction of $Cu(II)$ to $Cu(0)$. This work was done by Nekrasov and Berezina[5] shortly after the ring-disc electrode had been developed by Frumkin's school. If the reduction of $Cu(II)$ to $Cu(0)$ is carried out in a solution containing a ligand, for instance Cl^-, which stabilizes the $Cu(I)$, then the polarization curve for the reduction contains two well defined waves, corresponding to the two processes

$$Cu\,(II) \xrightarrow{e} Cu\,(I),$$

$$Cu\,(I) \xrightarrow{e} Cu\,(0).$$

Using a ring-disc electrode we can detect the $Cu(I)$ at the ring by oxidizing it back to $Cu(II)$:

    disc $Cu\,(II) \xrightarrow{e} Cu\,(I),$

                        $Cu\,(I) \xrightarrow{e} Cu\,(0),$

    ring $Cu\,(I) \xrightarrow{e} Cu\,(II).$

When the disc is at a potential that does not reduce the $Cu(I)$ then $|i_R| = N_0 |i_D|$, but as the second process takes over the ring current falls to zero. This is shown in Fig. 6.2.

If the reaction on the disc is carried out in a solution that does not contain a ligand to stabilize the $Cu(I)$, for example $CuSO_4$ and Na_2SO_4, then one wave rather than two waves are observed on the disc. It is then difficult to say from the shape of the polarization curve whether the reduction is taking place through $Cu(I)$ or not. However, a definite current due to $Cu(I)$ can be detected on the ring electrode as shown in Fig. 6.3. The maximum in the ring current occurs at the

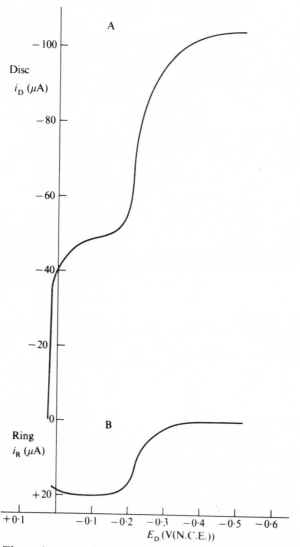

FIG. 6.2. The reduction of copper at a ring-disc electrode. Curve A: reduction of Cu^{2+} on a Au/Hg disc electrode. Curve B: corresponding limiting Cu^+ oxidation current on Au ring electrode. $[CuSO_4]$ 0·50 mM; $[KCl]$ 1M; $W = 86·7$ Hz. (After L.N. Nekrasov and N.P. Berezina (1962).)

foot of the disc wave since at more reducing potentials the disc removes all of the Cu (I).

More recently Miller[6] has carried out an interesting study of the anodic dissolution of Cu in 1 M NaOH using the split ring-disc electrode,

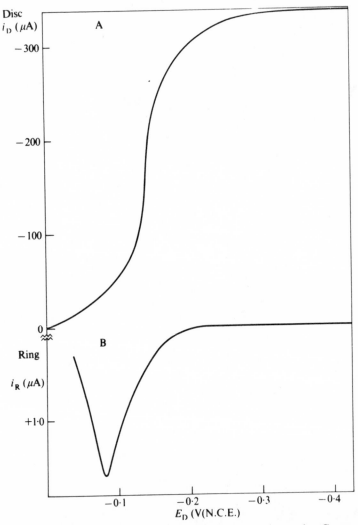

FIG. 6.3. The reduction of copper at a ring-disc electrode. Curve A: reduction of Cu^{2+} on a Pt disc electrode. Curve B: corresponding limiting Cu^+ oxidation current on the ring electrode. [$CuSO_4$] 30 mM; [Na_2SO_4] 1 M; $W = 35.9$ Hz. Note different scales for anodic and cathodic currents. (After L.N. Nekrasov and N.P. Berezina (1962).)

with the ring electrodes made of gold. Fig. 6.4 shows the results of sweeping the potential of a rotating copper electrode from a negative potential at a rate of $20\,\text{mV}\,\text{s}^{-1}$. Genuine steady-state currents cannot be observed in this system because of the growth of oxide layers on

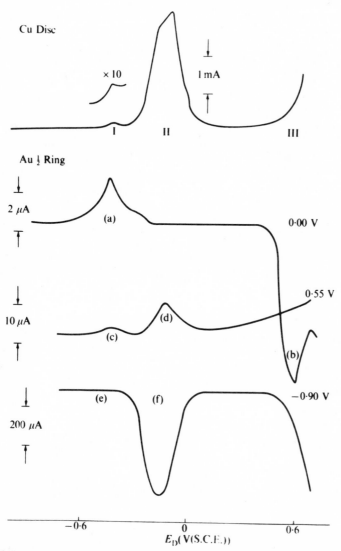

FIG. 6.4. Split ring-disc study of the anodic processes at a Cu electrode in alkaline solution. Traces from top: 1. Potentiodynamic sweep of Cu disc in 1 M NaOH. Au half-ring held at indicated potentials during disc sweep: 2. 0·00 V; 3. +0·55 V; 4. −0·90 V. (From Ref. 6, by permission of the Electrochemical Society.)

the electrode which passivate it and prevent further dissolution. The top trace for the disc shows three regions where current flows. Beyond region III oxygen evolution sets in. The three lower traces (on different scales) show the currents at the ring electrode, set at different potentials.

The electrode processes taking place may be listed:

(a) $Cu(I) \longrightarrow Cu(II)$,
(b) $Cu(III) \longrightarrow Cu(II)$,
(c) $Cu(I) \longrightarrow Cu(III)$,
(d) $Cu(II) \longrightarrow Cu(III)$,
(e) $Cu(I) \longrightarrow Cu(0)$,
(f) $Cu(II) \longrightarrow Cu(0)$.

The bump at e cannot be seen since the scale has to be made less sensitive to accommodate f. The identification of these processes was done by the split-ring technique. The disc is galvanostatted to pass a small current. If this process is started in region I and the two half-rings are potentiostatted at -0.7 V and $+0.3$ V (with respect to the S.C.E.) then they respectively reduce and oxidize the intermediate. The results are shown in Fig. 6.5. The mirror symmetry of the curves shows that the two half-ring processes must be

$$-0.7\,V \qquad Cu(I) \longrightarrow Cu(0);$$
$$+0.3\,V \qquad Cu(I) \longrightarrow Cu(II),$$

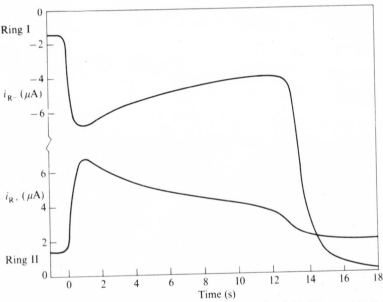

FIG. 6.5. Split ring anodic (i_{R+}) and cathodic (i_{R-}) limiting currents for copper anodisation in 1 M NaOH. $i_D = 80\,\mu A$; $E_{R+} = 0.3\,V$; $E_{R-} = -0.7\,V$; $W = 35\,Hz$. (From Ref. 6, by permission of the Electrochemical Society.)

and that the intermediate is therefore $Cu(I)$. Beyond 14 s the electrode
is covered with a Cu_2O layer, and to maintain the disc current (the
disc is galvanostatted) the disc potential becomes more oxidizing;
$Cu(II)$ is now generated on the disc (region II in Fig. 6.4). The reactions
on the half-rings become

$$-0.7\,V \qquad Cu(II) \longrightarrow Cu(0),$$
$$+0.3\,V \qquad\qquad residual.$$

Thus one current decays to zero, and the other current rises until it is
$N_0\,i_D$. From the coulombs supplied as disc current and the coulombs
measured at the ring the thickness of the Cu_2O layer can be calculated.

A similar study could be made of region III. This is complicated
by simultaneous oxygen evolution. However, by setting the two half-
rings at $-0.7\,V$ and $+0.2\,V$ and galvanostatting the disc in region III
it was found that the two half-ring currents obeyed the relation

$$i_{R,-0.7V} = (3.0 \pm 0.5)\,i_{R,+0.2V},$$

which would fit the two electrode reactions

$$+0.2\,V \qquad Cu(III) \longrightarrow Cu(II),$$
$$-0.7\,V \qquad Cu(III) \longrightarrow Cu(0).$$

Using these techniques further studies could be made of the $Cu(II)/$
$Cu(III)$ couple.

The Indium System

Miller and Visco[3] have also investigated the anodic dissolution
of indium in $0.7\,M\ HClO_4$. At low current densities linear i_R vs. i_D
plots were obtained for which

$$i_R = 2N_0 i_D.$$

Both currents were oxidizing currents; the potential of the ring
electrode was $+0.8\,V$ (vs. SCE) while the potential of the disc was
$\sim -0.5\,V$.

Using the split-ring technique, Miller and Visco found that with
the two half-rings potentiostatted at $+0.8\,V$ and $-0.8\,V$

$$i_{R,+0.8V} = -2\,i_{R,-0.8V}.$$

These two items of data show that the species being produced by the
disc electrode is $In(I)$ and the processes are

D

$$\text{disc} \qquad\qquad \ln(0) \xrightarrow{-e} \ln(I);$$

$$\text{ring} +0.8\,V \qquad \ln(I) \xrightarrow{-2e} \ln(III);$$

$$\text{ring} -0.8\,V \qquad \ln(I) \xrightarrow{e} \ln(0).$$

This agrees with the Tafel behaviour of the disc electrode. The $\ln(I)$ escapes into the solution where at low concentrations it reacts with H^+:

$$2\,H^+ + \ln(I) \longrightarrow \ln(III) + H_2.$$

At high concentrations of $\ln(I)$ a disproportionation reaction to $\ln(III)$ and $\ln(0)$ may also be a possibility.

At higher current density ($\sim 100\,mA\,cm^{-2}$) the observed collection efficiency falls off; that is

$$i_R < 2\,N_o\,i_D.$$

With the split ring

$$i_{R+1.0V} < 2\,N_{\frac{1}{2}}\,i_D$$

and

$$i_{R-0.65V} > N_{\frac{1}{2}}\,i_D.$$

These observations are consistent with the formation of some $\ln(II)$, but other explanations may also be possible.

The Oxygen Reaction

Damjanovic[7] has recently reviewed the mechanism of the electrochemical reduction of oxygen:

$$O_2 + 4\,H^+ \xrightarrow{4e} 2\,H_2O.$$

The ring-disc electrode has been used to detect and measure H_2O_2 formed either as an intermediate in the reaction, or as a by-product.

Some of the early results of Frumkin, Nekrasov, and Myuller[8,9] are shown in Fig. 6.6. The ring current is for the oxidation of the hydrogen peroxide.

For the reduction of oxygen on platinum in alkali the disc current polarization curves had a rather peculiar shape. There was no proper transport limited plateau but instead the disc current passed through a maximum and then decreased. The maximum was sharper the higher the rotation speed. Nekrasov and Myuller[9] explained these peculiar shapes by correcting them for the H_2O_2 detected on the ring. The basic scheme is disc $\quad O_2 + 2H_2O \xrightarrow{2e} H_2O_2 + 2OH^-;$

$H_2O_2 \xrightarrow{2e} 2OH^-;$

ring $\quad H_2O_2 \xrightarrow{-2e} O_2.$

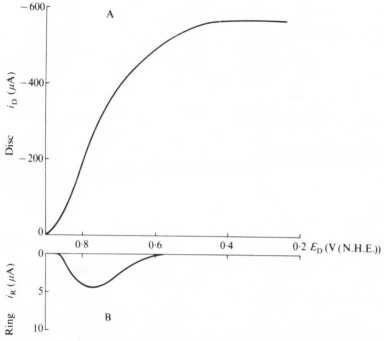

FIG. 6.6. The reduction of oxygen at a ring-disc electrode. Curve A: reduction of O_2 on a smooth Pt disc electrode. Curve B: corresponding limiting current for H_2O_2 oxidation on the ring electrode. [KOH] 0·125 M; $W = 5$ Hz. (After L.N. Nekrasov and L. Myuller (1963).)

Fig. 6.7 shows the disc and ring currents in the region of the transport limited current. Since the formation of H_2O_2 uses 2e per O_2 while the formation of H_2O uses 4e, the more H_2O_2 that is formed the smaller will be the total current per oxygen arriving at the disc electrode. Adding i_R/N_o to the observed disc current corrects it for this and leads to the calculated curves shown, which have the usual shape. Nekrasov and Myuller [9,10] have shown the increased production of H_2O_2 at potentials more negative than 0·5 V (with respect to the N.H.E.) is sensitive to the pre-treatment of the electrode and the extent of oxidation of its surface. They assume that the H_2O_2 is less stable on an oxidized than on a reduced surface and have suggested that H_2O_2 decomposes chemically without electron transfers to form

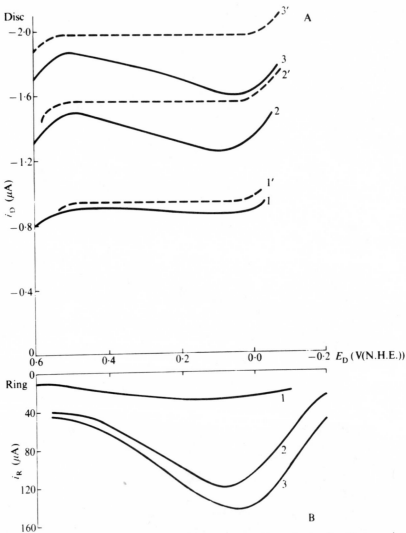

FIG. 6.7. The reduction of oxygen at a ring-disc electrode. Curves A: reduction of O_2 on a Pt disc in 0·125 M KOH. Curves B: corresponding limiting currents for H_2O_2 oxidation on the ring. W (Hz) (1) 17·33; (2) 49·17; (3) 76·00. 1′, 2′, and 3′ are disc polarization curves corrected for the amount of evolved H_2O_2. (After L.N. Nekrasov and L. Myuller (1963).)

O_2, which may then react again on the electrode. As the potential becomes more negative than 0·5 V the surface becomes more reduced, less H_2O_2 is decomposed back to O_2, and more reaches the ring electrode.

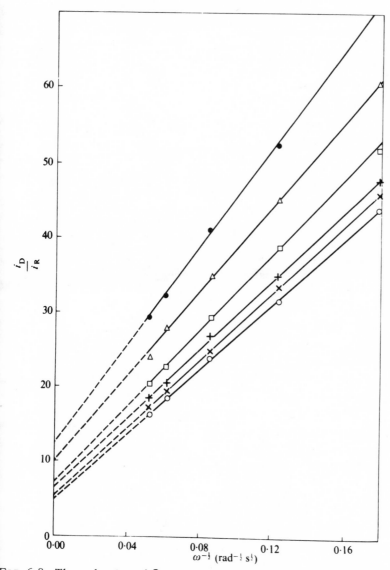

FIG. 6.8. The reduction of O_2 in alkaline solution at a ring-disc electrode – plot of i_D/i_R vs. $\omega^{-1/2}$ [eqn (6.7)]. [KOH] 0·1 M. Electrode is prereduced. E_D [V (N.H.E.)]: × 0·35; o 0·40; + 0·45; □ 0·50; Δ 0·55; ● 0·60. (From Ref. 11, by permission of the Electrochemical Society.)

Damjanovic, Genshaw, and Bockris[11] have also studied the reaction on platinum in alkali. They agree with Nekrasov and Myuller on the importance of the nature of the electrode surface. They suggested and have used the diagnostic plot discussed above. Fig. 6.8 shows some

results, and since both the intercepts and gradients vary with potential
the reaction scheme must be

$$O_2 \longrightarrow H_2O,$$

$$\dot{O}_2 \longrightarrow H_2O_2 \xrightarrow{k'} ?$$

From eqn (6.6)

$$x = \frac{\text{flux of } O_2 \longrightarrow H_2O}{\text{flux of } O_2 \longrightarrow H_2O_2}$$

and can be determined from the intercepts. It is found that x increases
as the potential decreases from 0.9 to ~ 0.6 V (vs. N.H.E.) passes
through a maximum and then decreases for potentials which are more
reducing than the 0.6 V. Analysis of the gradients shows that k' is
not very dependent on the electrode potential. It changes by about two
powers of 10 for a change of 0.9 V at the electrode. This confirms the
Russian workers' suggestion that k' is a chemical step that does not
involve electron transfer. The values obtained by three groups of
workers[11–13] are in reasonable agreement.

The greater production of H_2O_2 at more negative potentials is
caused by

(1) the fact that H_2O_2 is not itself decomposed electrochemically
as the potential is made more negative;

(2) x increasing in this region producing a higher proportion of
H_2O_2 for the available O_2.

Damjanovic et al.[11] have analysed the Tafel slopes and pH dependence
of the steps involving O_2 and have made suggestions about the
mechanism. For the route to H_2O_2 they suggest either

$$O_2 \underset{}{\overset{e}{\rightleftharpoons}} O_2^{\cdot -}$$

$$O_2^{\cdot -} + H_2O \longrightarrow HO_2^- + OH^{\cdot},$$

$$OH^{\cdot} \underset{}{\overset{e}{\rightleftharpoons}} OH^-;$$

or

$$O_2 + H_2O \underset{}{\overset{e}{\rightleftharpoons}} HO_2^{\cdot} + OH^-$$

$$HO_2^{\cdot} + H_2O \longrightarrow H_2O_2 + OH^{\cdot},$$

$$OH^{\cdot} \underset{}{\overset{e}{\rightleftharpoons}} OH^-.$$

It may be possible, however, that this is really a scheme of square
situation, for example,

$$O_2 \underset{\longleftarrow}{\overset{e}{\rightleftharpoons}} O_2^{\cdot -}$$

$$\downarrow$$

$$HO_2^{\cdot} \xrightarrow{\ e\ } HO_2^{-}$$

$$\downarrow$$

$$H_2O_2 \ .$$

The advantage of this type of scheme is that we do not involve $OH^{\cdot}$; it would seem likely that the reaction

$$OH^{\cdot} + HO_2^{-} \longrightarrow OH^{-} + HO_2^{\cdot}$$

would be downhill.

There is still controversy about the fate of the H_2O_2. Myuller and Nekrasov[10] prefer a catalysed disproportionation:

$$Pt + H_2O_2 \longrightarrow PtO + H_2O$$
$$\underline{PtO + H_2O_2 \longrightarrow Pt + H_2O + O_2}$$
$$2H_2O_2 \rightleftharpoons 2H_2O + O_2 \ .$$

They have studied the reduction of H_2O_2 on platinum under the same conditions.[10] A peculiar polarization wave is found (see Fig. 6.9). The reduction starts at the same potential as the oxygen wave when the electrode begins to decompose the oxygen produced from the disproportionation reaction. The current passes through a maximum at ~ 0.7 V (vs. N.H.E.) and then decreases gently to a minimum at 0.1 V — this decrease being caused by the removal of the PtO. Beyond 0.1 V the direct electrochemical reduction starts.

Damjanovic *et al.*[11] prefer either the rather formidable first step of

$$H_2O_2 \longrightarrow 2OH^{\cdot},$$

followed by

$$OH^{\cdot} \xrightarrow{\ e\ } OH^{-},$$

or

$$Pt + H_2O_2 \longrightarrow PtO + H_2O,$$

the same step as Myuller and Nekrasov, and then

$$PtO + H_2O \xrightarrow{\ 2e\ } Pt + 2OH^{-}.$$

The difference between this last scheme and that of the Russians is that the latter regenerate $\frac{1}{2}O_2$ for each H_2O_2. It is worth pointing out that this regeneration of the initial reagent does not invalidate the treatment of ring-disc data set out in eqn (6.7). The mathematics is not affected by the fate of B.

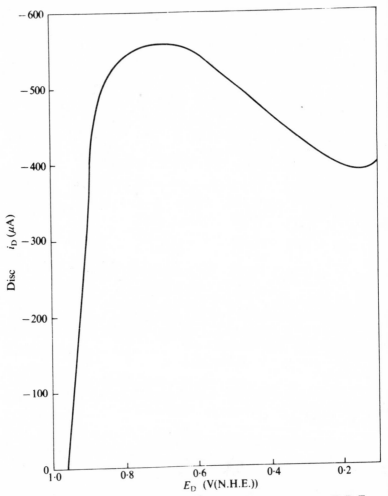

FIG. 6.9. The reduction of H_2O_2 in alkaline solution at an R.D.E. (After L. Myuller and L.N. Nekrasov (1964).)

Many other studies have been done with the ring-disc electrode on the oxygen reaction. For instance, Damjanovic et al.[14] have obtained the plots of eqn (6.7), shown in Fig. 6.10, for the reduction at Pt in acid solution.

From Table 6.1 the zero gradients but displaced lines show that H_2O_2 is formed as a minor side product and is not itself further consumed on the electrode. Similar studies have also been made at electrodes of other materials, for example, rhodium,[15,16] palladium,[17] silver,[18] and gold.[19]

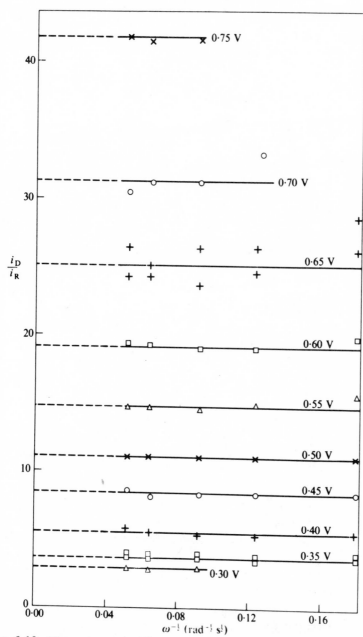

FIG. 6.10. The reduction of O_2 in acid solution at a ring-disc electrode
– plot of i_D/i_R vs. $\omega^{-1/2}$ [eqn (6.7)]. [H_2SO_4] 0.05 M. E_D values are
given in the figure. (From Ref. 14, by permission of the Electrochemical
Society.)

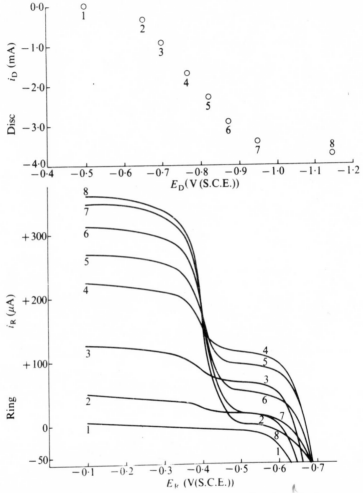

FIG. 6.11. Ring electrode polarization curves for the nitrobenzene system. [NaOH] 0·10 M. The top figure shows the points on the disc polarization curve at which the corresponding ring polarization curves were measured. Note that the disc polarization curve does not consist of two waves but that the ring electrode does detect two products.

Reduction of Nitro benzene

The authors have been using a gold ring-disc electrode to investigate the reduction of nitrobenzene in alkaline solutions. The reduction proceeds as far as phenylhydroxylamine, which means that four electrons have been added. Thus the scheme of squares is 5 by 5, and

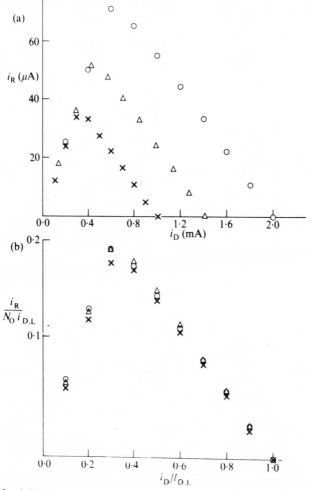

FIG. 6.12. (a) Ring current as a function of disc current. (b) Normalized ring current vs. normalized disc current. W (Hz) × 5; Δ 10; o 20. NaOH] = 3 M; $E_R = -0.565$ V.

herefore the elucidation of the complete mechanism is a formidable ask, which is not yet completed. Two products can be detected when he ring electrode is at oxidizing potentials. One product is phenyl-ydroxylamine and this is oxidized back to nitrosobenzene at potentials more positive than -0.42 V (with respect to the S.C.E.). This was hown by using known samples of phenylhydroxylamine and nitroso-enzene. At low disc currents a second product could be oxidised at ower potentials. Fig. 6.11 shows some typical ring polarization curves

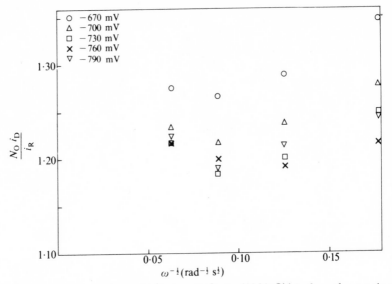

FIG. 6.13. Plots of eqn (6.7) for ϕNO_2 in IM NaOH before the maximum. Note the expanded scale on the y-axis.

as a function of disc current. The ring current with the ring electrode potentiostatted at -565 mV is shown in Fig. 6.12 as a function of disc current. It can be seen to rise to a maximum for lowish disc currents and then to decay to nothing as the disc is made more reducing. The maximum is not placed symmetrically but occurs roughly at one-quarter of the limiting disc current. This is particularly true for solutions in the strongest alkali (~ 1 to $3\,M$); as the solutions become more acid the ring current curve collapses and the maximum shifts to a more symmetrical position. Experiments with nitrosobenzene show that it is easily reduced at potentials which reduce nitrobenzene. This fact together with the $1:3$ ratio suggest that the intermediate is the nitrobenzene radical anion. This species has been generated electrochemically and studied by E.S.R.[20]

We found that the $i_R - i_D$ plots could be normalized on to a common curve by dividing both i_R and i_D by the limiting disc current. Fig. 6.12 also shows the normalization of a family of curves. This indicates that the nitrobenzene radical anion is not decomposing homogeneously in its passage from the disc to the ring. As discussed in subsequent chapters N_K for homogeneous kinetics is invariably rotation speed dependent.

Different behaviour is found for the rise to and fall from the maximum. Figs. 6.13 and 6.14 show plots of eqn (6.7) for the rise and fall respectively at different values of pH. On the way up the collection

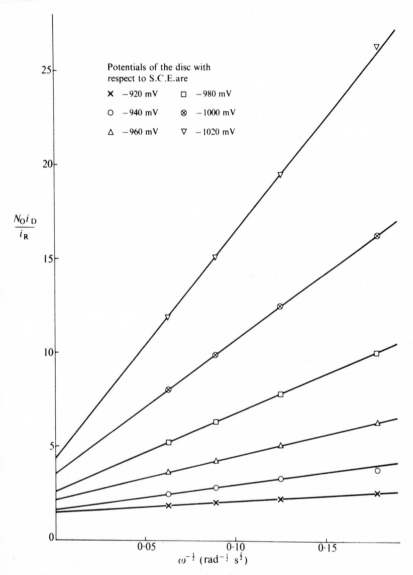

FIG. 6.14. Plots of eqn (6.7) for ϕNO_2 in IM NaOH after the maximum.

efficiency is not much affected by rotation speed. This can also be seen from the normalization in Fig. 6.12 where whatever the rotation speed i_R is a constant fraction of i_D. Considering that the potential changes by ~ 200 mV the fraction is relatively independent of potential. We can conclude that in this region the radical anion seen on the ring electrode is not destroyed on the disc and is not an intermediate in the production of the other products, which is mainly ϕNHOH. This has been confirmed, using the split-ring technique to measure at any instant how much of the disc current is producing $\phi NO_2^{\cdot-}$ and how much ϕNHOH.

The mechanism has the general form

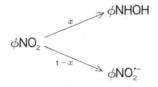

The fact that x is independent of potential shows that in the branching of the mechanism the competition must be between two chemical steps or between two similar electron transfer steps. The effect of decreasing the pH is to decrease x. However, x is also sensitive to the pre-treatment of the electrode and at the moment no definite conclusions can be drawn about the order of x with respect to $[OH^-]$. Beyond the maximum x depends upon potential. The more reducing the electrode the larger is x. If at various pHs a Tafel plot is made of $\log x$ against potential the slopes are found to lie in the range 120–170 mV. Similar slopes are found for plots of $\log k'$ calculated from eqn (6.7), and indeed in any experiment the potential dependence of $\log k'$ and of $\log x$ are found to be similar. The value of 170 mV corresponds to an α of 0·28 and this seems too small to be a genuine one-electron transfer.

The existence of rotation speed independent branching in the mechanism is also somewhat surprising since it is not easy to devise a route for the production of ϕNHOH which does not pass through $\phi NO_2^{\cdot-}$. It is possible that the desorption of $\phi NO_2^{\cdot-}$ from the electrode is rate determining or the results may be explained by a two-site model of the gold electrode, which assumes an electrode mechanism of the type

$$
\phi NO_2 \Bigg\langle
\begin{array}{l}
\nearrow \ \phi NO_2' \longrightarrow \phi NO_2^{\cdot-}{}' \longrightarrow \phi NO \longrightarrow \phi NHOH \\
\hspace{3.5cm} \uparrow \\
\hspace{3.5cm} | \\
\searrow \ \phi NO_2 \longrightarrow \phi NO_2^{\cdot-} \leadsto \text{solution and ring electrode.}
\end{array}
$$

($\leadsto$ are diffusive steps)

The primed species are on special sites which remove oxygen from the $\phi NO_2^{\cdot-}$ to form ϕNO. We observe that less $\phi NO_2^{\cdot-}$ is formed at more reducing potentials, at more acid pHs, and if the electrode is left at a reducing potential of $-1\cdot150$ V before starting a sweep. These facts could all be explained by postulating that the breaking of the O–N bond in $\phi NO_2^{\cdot-}$ is catalysed by a particular site on the electrode. This work is as yet incomplete and so this mechanism can only be regarded as a tentative suggestion.

Conclusions

In conclusion we summarize some of the uses of the ring-disc electrode in the study of electrode mechanisms.

(1) To detect intermediates.

(2) To characterize intermediates either by the half wave potential or by their oxidation state.

(3) To measure the rates of decomposition of the intermediates on the disc electrode.

(4) To characterize the pH profile of this rate.

(5) To measure directly the surface concentration of the intermediate on the disc electrode.

(6) To measure the ratio of different products produced by the electrode.

References

1. IVANOV, Y.B. and LEVICH, V.G. (1959) *Dokl. Akad. Nauk SSSR* 126, 1029.

2. DAMJANOVIC, A., GENSHAW, M.A., and BOCKRIS, J.O'M. (1966) *J. chem. Phys.* 45, 4057.

3. MILLER, B. and VISCO, R.E. (1968) *J. electrochem. Soc.* 115, 251.

4. ALBERY, W.J. and WORMALD, E.M. (1969) Unpublished results.

5. NEKRASOV, L.N. and BEREZINA, N.P. (1962) *Dokl. Akad. Nauk SSSR* 142, 855.

6. MILLER, B. (1969) *J. electrochem. Soc.* 116, 1675.

7. DAMJANOVIC, A. (1969) In *Modern aspects of electrochemistry* (eds. J.O'M. Bockris and B.E. Conway), Vol. 5. Butterworths.

8. FRUMKIN, A.N. and NEKRASOV, L.N. (1959) *Dokl. Akad. Nauk SSSR* 126, 115.

9. NEKRASOV, L.N. and MYULLER, L. (1963) *Dokl. Akad. Nauk SSSR* 149, 1107.

10. MYULLER, L. and NEKRASOV, L.N. (1964) *Dokl. Akad. Nauk SSSR* 157, 416.

11. DAMJANOVIC, A., GENSHAW, M.A., and BOCKRIS, J.O'M. (1967) *J. electrochem. Soc.* 114, 1107.

12. MYULLER, L. and NEKRASOV, L.N. (1965) *J. electroanalyt. Chem.* 9, 282.

13. JACQ, J. and BLOCH, O. (1964) *Electrochim. Acta* 9, 551.

14. DAMJANOVIC, A., GENSHAW, M.A., and BOCKRIS, J.O'M, (1967) *J. electrochem. Soc.* 114, 466.

15. NEKRASOV, L.N., KHRUSHCHEVA, E.I., SHUMILOVA, N.A., and TARASEVICH, M.R. (1966) *Elektrokhimya* 2, 363.

16. GENSHAW, M.A., DAMJANOVIC, A., and BOCKRIS, J.O'M. (1967) *J. phys. Chem. Ithaca* 71, 3722.

17. SOBOL, V.V., KHRUSHCHEVA, E.I., and DAGAEVA, V.A. (1965) *Elektrokhimya* 1, 1332.

18. ZHUTAEVA, G.V., SHUMILOVA, N.A., and TARASEVICH, M.R. (1965) *Dokl. Akad. Nauk SSSR* 161, 151.

19. GENSHAW, M.A., DAMJANOVIC, A., and BOCKRIS, J.O'M. (1967) *J. electroanalyt. Chem.* 15, 163.

20. KOOPMANN, R. and GERISCHER, H. (1966) *Ber. Bunsenges* 70, 128.

7

TITRATION CURVES

Description of Titration Curve

WHEN the intermediate generated on the disc electrode is rapidly
destroyed by some other species in the solution then the shape of the
ring current - disc current plot has the form shown in Fig. 7.1.

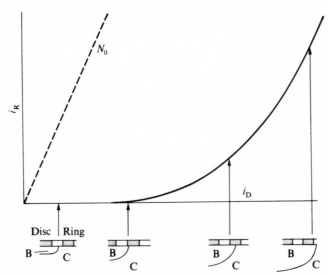

FIG. 7.1. Ring-disc electrode titration curve.

The system that we are discussing has the general form

$$\text{disc electrode} \qquad A \longrightarrow B\,;$$

$$\text{solution} \qquad B + C \xrightarrow{k_2} \text{products;} \qquad \text{fast reaction}$$

$$\text{ring electrode} \qquad B \longrightarrow A.$$

An example would be

$$A = Br^-; \quad B = Br_2; \quad \text{and} \quad C = As(III)$$

If the reaction between B and C is very fast then no B can exist in the presence of C and vice versa. The reaction takes place at an interzonal boundary between the B-dominated region and the C-dominated region. There can be no B-dominated region until the disc current generating B produces a flux of B which is larger than the transport limited flux of C to the disc electrode. This critical disc current is given the symbol M. When $|i_D| = M$, the electrochemical generation of B exactly matches the pumping action of the disc which brings C from the bulk of the solution to the reaction boundary on the electrode surface. Since the concentration of C is then zero at the electrode surface it has the same concentration profile as shown in Fig. 2.5 for a transport limited current. For $|i_D| > M$ there will be the B-dominated region of the solution around the disc electrode. The larger is i_D the further this region will extend into the diffusion layer; the bulk of the solution always remains dominated by C. No ring current can be observed until the B-dominated region has reached the inside edge of the ring electrode. This occurs when:[1]

$$|i_D| = M/[1 - F(\alpha)], \tag{7.1}$$

where

$$\alpha = (r_2/r_1)^3 - 1$$

and α describes the thickness of the gap.

As the B region moves further across the ring electrode the ring current rises and eventually when the ring electrode lies entirely in the B-dominated region the ring current becomes a linear function of the disc current:

$$i_R = N_0|i_D| - M\beta^{2/3}, \tag{7.2}$$

where

$$\beta = (r_3/r_1)^3 - (r_2/r_1)^3.$$

Fig. 7.2 shows a typical concentration profile for both B and C when the reaction boundary is on the ring electrode. Fig. 7.1 shows a strip cartoon of the titration curve and the spreading out of the B-dominated region.

Calculation of Titration Curve

The shape of this curve can be calculated theoretically.[1] The concentration profiles of both B and C must be described and we must therefore write

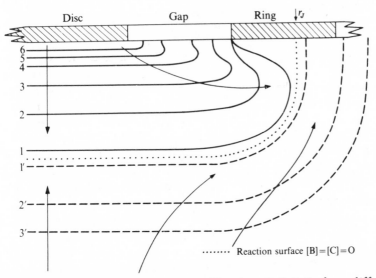

FIG. 7.2. Schematic concentration profile at an R.R.D.E. for a diffusion layer titration. Reaction surface $[B] = [C] = 0$.

$$v_z \frac{\partial b}{\partial z} + v_r \frac{\partial b}{\partial r} = D_B \frac{\partial^2 b}{\partial z^2} - k_2 bc; \tag{7.3}$$

$$v_z \frac{\partial c}{\partial z} + v_r \frac{\partial c}{\partial r} = D_C \frac{\partial^2 c}{\partial z^2} - k_2 bc. \tag{7.4}$$

The difference between these equations and eqn (3.6) is the term $-k_2 bc$ describing the irreversible reaction between B and C. If we assume that $D_B = D_C$ then subtraction of the two equations removes the nasty non-linear term and gives

$$v_z \frac{\partial u}{\partial z} + v_r \frac{\partial u}{\partial r} = D \frac{\partial^2 u}{\partial z^2} ,$$

where

$$u = \frac{\pi r_1^2 n_D F D^{2/3} C^{1/3}}{|i_D|} (c_\infty - c + b).$$

The function u has been chosen so that it satisfies the three boundary conditions set out in eqn (3.8) and, using the variables in eqn (3.7), the differential equation becomes

$$\frac{\partial u}{\partial \xi_n} = \frac{1}{y} \frac{\partial^2 u}{\partial y^2} .$$

The radial position of the reaction boundary is important in determining the ring current. The boundary is defined as the surface where $b = c$ and

$$u = \frac{\pi r_1^2 n_D FD^{2/3} C^{1/3}}{|i_D|} c_\infty = \frac{A_1 M}{|i_D|},$$

where

$$M = \frac{\pi r_1^2 n_D FD^{2/3} C^{1/3} c_\infty}{A_1}. \tag{7.5}$$

Let its radial distance on the surface of the disc be r_J. Then for

$$r_J < r_2, \quad i_R = 0,$$

since no B is reaching the ring electrode.

For $r_2 < r_J < r_3$ the boundary condition for u on the ring electrode is given by

$$r_2 < r < r_J, \quad u_0 = \frac{A_1 M}{|i_D|} \tag{7.6}$$

since $c = 0$ in the B-dominated region and $b = 0$ because it is being destroyed on the ring electrode, and for

$$r_J < r, \quad \left(\frac{\partial u}{\partial y}\right)_0 = 0, \tag{7.7}$$

since $b \sim 0$ in the C-dominated region, and $\partial c/\partial y = 0$ on the ring electrode since it does not react. The value of r_J is the point where both boundary conditions (7.6) and (7.7) hold. The ring current is given by

$$|i_R| = 2\pi n_R FD \int_{r_2}^{r_J} (\partial b/\partial z)_0 \, r \, dr, \tag{7.8}$$

or

$$|i_R/i_D| = 2 \int_0^{\beta_J} (\partial u/\partial y)_0 \, d\xi_2, \tag{7.9}$$

where

$$\beta_J = (r_J/r_1)^3 - (r_2/r_1)^3.$$

Solution of the differential equation with its boundary conditions and the satisfaction of both boundary conditions (7.6) and (7.7) at $r = r_J$ gives [1]

$$N' = |i_R/i_D| = 1 - F\left(\frac{\alpha}{\beta_J}\right) - \frac{1+\alpha}{(1+\alpha+\beta_J)^{1/3}} \left[1 - F\left\{\frac{(1+\alpha+\beta_J)^\alpha}{\beta_J}\right\}\right] \tag{7.10}$$

and

$$\frac{M}{|i_D|} = 1 - F(\alpha) - \frac{\beta_J^{1/3}}{(1+\alpha+\beta_J)^{1/3}} \left[1 - F\left\{\frac{(1+\alpha+\beta_J)^\alpha}{\beta_J}\right\}\right]. \tag{7.11}$$

These two equations describe pairs of values of i_R and i_D for each value of β_J. In particular at $\beta_J = 0$, $r_J = r_2$, we obtain eqn (7.1).

When $r_J > r_3$, then boundary condition (7.6) holds for:

$$r_2 < r < r_3.$$

r_J is replaced by r_3 in eqn (7.8) and β_J is replaced by β in equations (7.9)–(7.11). Since from eqn (3.12)

$$N_0 = 1 - F(\alpha/\beta) + \beta^{2/3}\{1 - F(\alpha)\} -$$
$$- (1 + \alpha + \beta)^{2/3}[1 - F\{(\alpha/\beta)(1 + \alpha + \beta)\}],$$

multiplication of (7.11) by $\beta^{2/3}$ and addition to eqn (7.10) gives:

$$N_0 = \frac{|i_R|}{|i_D|} + \frac{\beta^{2/3}M}{|i_D|}$$

or

$$|i_R| = N_0|i_D| - \beta^{2/3}M, \tag{7.12}$$

as given above in eqn (7.2).

This equation describes the displacement of the titration curve from the straight line:

$$|i_R| = N_0|i_D|,$$

which is observed when $M = 0$. The quantity M, given in eqn (7.5), is proportional to c_∞, the bulk concentration of the reagent which reacts with the intermediate. It is also proportional to $C^{1/3}$ which means that it is proportional to $\omega^{1/2}$. The reason for this is that the faster the electrode is turned, the more the reagent is pumped up to the region of the disc electrode.

Analysis of Experimental Results

The simplest way of analysing a titration curve is to make use of this limiting eqn (7.12).[2] The straight part of the curve is extrapolated back to the point where $i_R = 0$, and at this point:

$$M = \frac{N_0|i_D|}{\beta^{2/3}}$$

and from eqn (7.5)

$$c_\infty = \frac{0 \cdot 205 \, \nu^{1/6} M}{r_1^2 n_D F D^{2/3} W^{1/2}} \tag{7.13}$$

or

$$c_\infty = \frac{0 \cdot 205\, N_0 \, |\, i_D \, |\, \nu^{1/6}}{\beta^{2/3} \, r_1^2 \, n_D F D^{2/3} \, W^{1/2}} \, ,$$

where W is the rotation speed measured in Hz. A problem with this simple procedure is that it is sometimes difficult to be certain where the titration curve straightens out and the extrapolation may be inaccurate.

The complete analysis of the titration curve can be done by using eqn (7.10) and (7.11). For each electrode values of β_J between 0 and β are taken and pairs of values of N' and $M/|\, i_D\, |$ are calculated. This calculation has only to be performed once for each electrode.

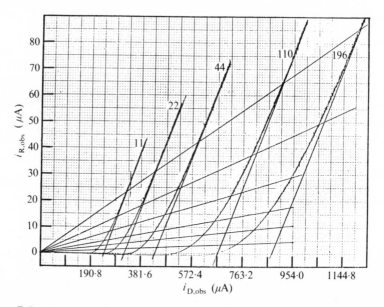

FIG. 7.3. Experimental diffusion layer titration curves for the $Br_2/As(III)$ system. $[As\,(III)] = 0 \cdot 527$ mM; $[KBr] = 0 \cdot 20$ M; $[H_2SO_4] = 1 \cdot 00$ M. Rotation speed for each curve given in rad s^{-1}.

Experimental i_R/i_D traces[1] for the Br_2–As (III) systems are shown in Fig. 7.3 for five different rotation speeds. The system is

disc electrode $2\,Br^- \longrightarrow Br_2$;

solution $Br_2 + As\,(III) \longrightarrow As\,(V) + 2\,Br^-$;

ring electrode $Br_2 \longrightarrow 2\,Br^-$.

The experimental i_R/i_D traces are analysed by drawing lines of gradient N' and reading off values of $i_{D,obs}$ where these constructed lines cross the curves. Each $i_{D, obs}$ is then plotted against the value of $|i_D|/M$ corresponding to that particular N', as shown in Fig. 7.4. The gradient of each line is M for that particular rotation speed. The deviation of the points at high rotation speeds and the low disc currents are caused by the finite rate of the reaction; the theory has assumed that the reaction is infinitely fast. This kinetic effect is discussed in the next chapter.

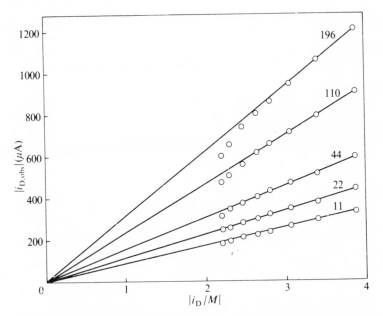

FIG. 7.4. Plot of observed disc current, $|i_{D, obs}|$, against $|i_D/M|$ — test of eqns (7.10) and (7.11). Rotation speeds given in rad s^{-1}.

Fig. 7.5 shows values of M plotted against the concentration of As (III) for the different rotation speeds, and Fig. 7.6 shows the dependence of the gradients from Fig. 7.5, $M/[As (III)]$, on $\omega^{1/2}$. The good straight line confirms the theory. The technique has been applied to the measurements of concentrations of As (III) as low as $\sim 10^{-7} M$.

In eqn (7.13), which relates c_∞ and M, it is necessary to know ν and D. Since ν is to the power of $1/6$ it does not need to be known too accurately.

The diffusion coefficient D may be determined by measuring M for a solution of known c_∞, or by a new ingenious technique. The disc is galvanostatted so that the reaction zone is beyond the ring electrode and eqn (7.12) applies:

$$i_R = N_0 |i_D| - \beta^{2/3} M$$
$$= N_0 |i_D| - \beta^{2/3} c_\infty nFM'.$$

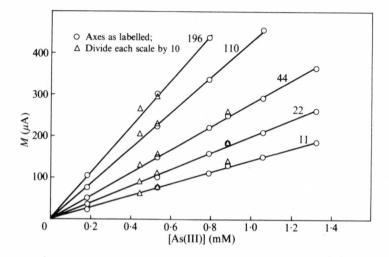

FIG. 7.5. Dependence of M on [As(III)]. O, Axes as labelled; △, Divide each scale by 10. Rotation speeds given in rad s^{-1}.

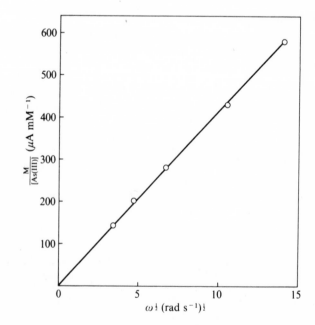

FIG. 7.6. Dependence of M/[As(III)] on $\omega^{1/2}$.

We can relate the depletion with time of the bulk concentration of C to the currents at the electrodes:

$$V\frac{\partial c_\infty}{\partial t} = -\frac{(|i_D| - |i_R|)}{nF},$$

where V is the volume of the solution.

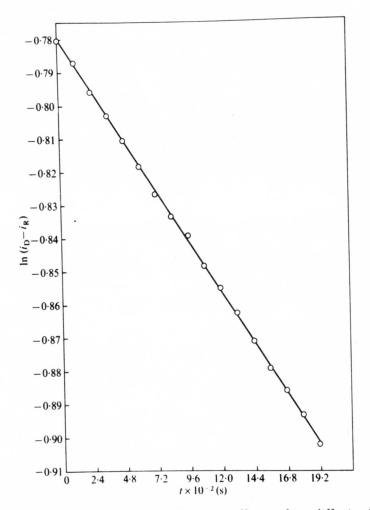

FIG. 7.7. Determination of the diffusion coefficient for a diffusion layer titration. System Br_2/As (III). $[As\,(III)]_{t=0} \sim 0.19\,mM$; $[KBr] = 0.20\,M$; $[H_2SO_4] = 1.00\,M$. From least squares slope and eqn (7.14): $D = 0.975 \times 10^{-5}\,cm^2\,s^{-1}$. From Ref. 1 and eqn (7.5): $D = 0.98 \times 10^{-5}\,cm^2\,s^{-1}$.

Then

$$\frac{\partial |i_R|}{\partial t} = \frac{(|i_D| - |i_R|)}{V} \beta^{2/3} M',$$

and

$$\ln(|i_D| - |i_R|) = \text{a constant} - \frac{\beta^{2/3} M'}{V} t. \qquad (7.14)$$

Hence M' and D can be found. Fig. 7.7 shows some typical results.[3] Thus all the necessary parameters for determining concentration can be found from analysis of the titration curve and its variation with time. There is no need to determine D or c_∞ by independent measurement.

In applying this technique to other systems it is not necessary for the species C to be inactive on the disc electrode, since when there is a finite ring current the disc is in the B-dominated region, and all the C will be destroyed before reaching it. It is necessary, however, for the reaction between B and C to be rapid ($> \sim 10^5 M^{-1} s^{-1}$), and for the products of this reaction not to interfere with the detection of B at the ring.

References

1. ALBERY, W.J., BRUCKENSTEIN, S., and JOHNSON, D.C. (1966) *Trans. Faraday Soc.* 62, 1938.

2. BRUCKENSTEIN, S. and JOHNSON, D.C. (1964) *Analyt. Chem.* 36, 2186.

3. ALBERY, W.J. and HITCHMAN, M.L. (1970) Unpublished results.

8

HOMOGENEOUS KINETICS – SECOND ORDER

Introduction

The ring-disc electrode is a useful technique for the measurement of
the kinetics of fast reactions taking place in the solution. In this
application it has some analogies with a flow system. The disc
electrode is used as a generator of one of the reactants; its rate of
generation is easily controlled by means of the disc current. The ring
electrode placed 'down stream' is the detector, and measures the
amount of electrogenerated reactant that has survived the passage
from the disc to the ring.

However, in this flow system the flow is laminar and the transport
can be calculated. Unlike stopped flow, it is a steady-state technique
that makes the experiments easier to carry out. Unlike continuous
flow methods, there is no need for a continuous supply of reactant
solutions, since one of the reagents is generated *in situ* on the disc
electrode.

The technique can be used under two different conditions, depending
on whether the electrogenerated intermediate is destroyed by first- or
second-order kinetics. For first-order kinetics the intermediate may
undergo a unimolecular decomposition, or it may react with the solvent,
or with a species in such a large excess that its concentration does
not vary. For second-order kinetics the intermediate reacts with a
species of comparable concentration to its own, and the concentration
profiles of both species have to be calculated. Since the theory of
second-order kinetics follows on from the titration curve it will be
treated first.

Qualitative Description

The system is the same as that of the titration curve, except that
the reaction between B and C is not assumed to be infinitely fast, but
is given the rate constant k_2:

$$\text{disc} \qquad A \longrightarrow B;$$

$$\text{solution} \qquad B + C \xrightarrow{k_2} \text{products};$$

$$\text{ring} \qquad B \longrightarrow A.$$

The differential equations are the same as eqns (7.3) and (7.4):

$$v_z \frac{\partial b}{\partial z} + v_r \frac{\partial b}{\partial r} = D_B \frac{\partial^2 b}{\partial z^2} - k_2 bc, \qquad (7.3)$$

$$v_z \frac{\partial c}{\partial z} + v_r \frac{\partial c}{\partial r} = D_C \frac{\partial^2 c}{\partial z^2} - k_2 bc. \qquad (7.4)$$

Following the same argument we can solve for u in exactly the same way as in the titration curve. In particular, from eqn (7.1) when

$$\frac{|i_D|}{M} = \frac{1}{1 - F(\alpha)},$$

the reaction boundary, where $b = c$ on the surface of the disc ($z = 0$) is placed on the inside edge of the ring electrode at $r = r_2$. For the infinitely fast reaction at this point b and c would both tend to zero and there would be no ring current. However, in fact, b and c both have the same finite value. Instead of there being a sharp reaction boundary there is a zone where some B penetrates into the C-dominated region and vice versa. This is shown in Fig. 8.1.

The amount of penetration of B into the C-dominated region is very dependent upon the rate of reaction between B and C. When the disc current is adjusted so that the reaction boundary ($b = c$) is placed on the inside edge of the ring electrode, then the ring current is entirely caused by the penetration of B into the C region, and this is the best conditions under which to measure the second-order rate constant.

Mathematical Theory

It is impossible to solve the non-linear differential eqns (7.3) and (7.4) analytically to obtain the ring current. However, we may first draw an important general conclusion. Normalizing eqns (7.3) and (7.4) with

$$c = g c_\infty, \qquad b = h c_\infty, \qquad w = z (C/D)^{1/3},$$

$$\xi_3 = \ln (r/r_2),$$

and

$$\kappa = \left(\frac{k_2 c_\infty}{D} \right)^{1/2} \left(\frac{D}{C} \right)^{1/3}, \qquad (8.1)$$

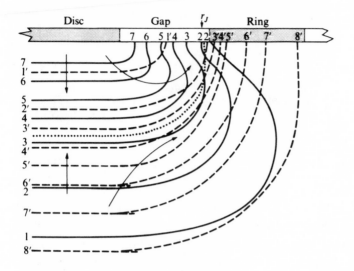

Interpenetration of reactants

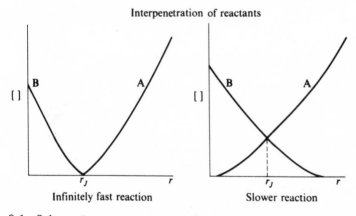

Infinitely fast reaction   Slower reaction

FIG. 8.1. Schematic concentration profile at a ring-disc electrode for the intermediate reacting with another species. ———— Intermediate B; -------- Reagent C; .... Surface of equal concentration.

we obtain

$$ w \frac{\partial g}{\partial \xi_3} - w^2 \frac{\partial g}{\partial w} = \frac{\partial^2 g}{\partial w^2} - \kappa^2 gh \tag{8.2} $$

and

$$ w \frac{\partial h}{\partial \xi_3} - w^2 \frac{\partial h}{\partial w} = \frac{\partial^2 g}{\partial w^2} - \kappa^2 gh. \tag{8.3} $$

The boundary conditions are

$$ w \longrightarrow \infty, \quad g \longrightarrow 1, \quad h \longrightarrow 0; $$

disc electrode

$$r < r_1, \quad w = 0, \quad \partial g/\partial w = 0, \quad \partial h/\partial w = -\frac{1}{A_1\{1 - F(\alpha)\}} ;$$

gap

$$r_1 < r < r_2, \quad w = 0, \quad \partial g/\partial w = 0, \quad \partial h/\partial w = 0;$$

ring electrode

$$r_2 < r < r_3, \quad w = 0, \quad \partial g/\partial w = 0, \quad h = 0.$$

The boundary condition for h on the disc electrode is equivalent to

$$\frac{|i_D|}{M} = \frac{1}{1 - F(\alpha)},$$

which places the reaction boundary on the inside edge of the ring electrode. We now define N_K', the kinetic collection efficiency, which is the ring current observed under these conditions, divided by the disc current:

$$N_K' = 2 A_1 \{1 - F(\alpha)\} r_1^{-2} \int_{r_2}^{r_3} (\partial h/\partial w)_{w=0} r \, d r.$$

Now for a particular electrode and for any system with the same value of κ the solution for $\partial h/\partial w$ from eqns (8.2) and (8.3) and the boundary conditions will be the same, and will give the same answer for N_K'. Thus we can conclude that N_K' is a function of κ and the electrode geometry only.

An approximate analytical solution for the ring current and N_K' may be obtained from eqns (8.2) and (8.3) using the method of moments.[1,2] Three assumptions have to be made. First, that the reaction is fast enough for the term describing normal convection to be negligible; that is the penetration in the vicinity of the ring electrode is close to the electrode surface. The condition for this[2] is that

$$Q' \ll 1, \tag{8.4}$$

where

$$Q' = 0.64 C^{2/5} D^{1/5} / (k_2 c_\infty P)^{3/5}$$

and

$$P = \frac{3^{3/2} r_1^2}{4[1 - F(\alpha)](r_2^3 - r_1^3)^{2/3}} \simeq 3.$$

For $D = 10^{-5} \text{ cm}^2 \text{ s}^{-1}$ and $\nu = 10^{-2} \text{ cm}^2 \text{ s}^{-1}$.

$$Q' \simeq 0.06 (\omega/k_2 c_\infty)^{3/5}. \tag{8.5}$$

The second assumption is that the ring electrode is wide enough to collect all the intermediate that penetrates along the disc surface.

The condition for this is[2]

$$\frac{r_3 - r_2}{r_2} > Q'. \qquad (8.6)$$

The third assumption is that on an inert surface at $r = r_2$ and $z = 0$

$$\frac{\partial b}{\partial r} = -\frac{\partial c}{\partial r}.$$

This assumes that the concentration profiles describing the penetration are symmetrical at the reaction boundary as shown in Fig. 8.1.

We then obtain[2]

$$i_{R,\kappa} = 0{\cdot}21 \, \pi \, r_1^2 n F D \omega^{3/2} \nu^{-1/2} k_2^{-1} \qquad (8.7)$$

and

$$N_\kappa' = \frac{0{\cdot}339 \, r_2^2 D^{1/3} \{1 - F(\alpha)\}}{r_1^2 \nu^{1/3} k_2} \cdot \frac{\omega}{c_\infty} \qquad (8.8)$$

The Reaction Between Bromine and Allyl Alcohol

These equations and technique have been tested by studying the bromination of allyl alcohol.[2] This reaction was chosen because its rate constant had been measured by Bell and Atkinson[3] using a low-concentration potentiometric technique. They found that the second-order rate constant for the reaction was about $2 \times 10^5 \, M^{-1} s^{-1}$. Concentrations of bromine as low as $\sim 10^{-7}$ M can be measured potentiometrically, and for such concentrations the half time for the disappearance of allyl alcohol can be as long as ~ 100 s. The reaction can therefore be followed classically by measuring the disappearance of Br_2 with time.

In using the ring-disc technique we obtain a ring current – disc current trace which has the shape of a titration curve (see Fig. 8.2). Since we wish to measure accurately the amount of penetration when the Br_2 first reaches the ring electrode, we amplify the ring current scale for the first part of the trace (see Fig. 8.3). The disc current $i_{D,\kappa}$, which corresponds to the reaction zone being on the inside edge of the ring electrode, is obtained from eqn (7.1):

$$|i_{D,\kappa}| = \frac{M}{1 - F(\alpha)};$$

M and $i_{D,\kappa}$ are determined experimentally from the full titration curve (Fig. 8.2) as described in the last chapter. Fig. 8.4 shows the $i_{D,obs}$ vs. i_D/M plot obtained from Fig. 8.2. The observed disc currents for

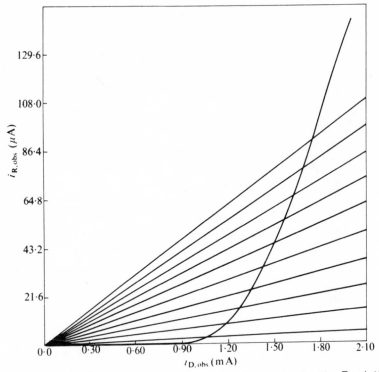

FIG. 8.2. Experimental diffusion layer titration curve for the Br_2/allyl alcohol system. [allyl alcohol] = 1·44 mM; [NaBr] = 0·20 M; W = 5 Hz. Lines of slope N' are drawn on the experimental curve.

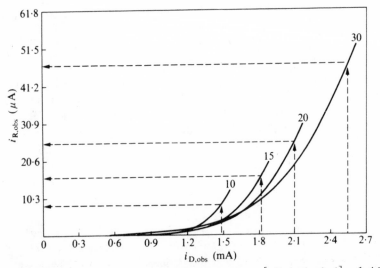

FIG. 8.3. Experimental $i_R - i_D$ kinetic curves. [allyl alcohol] = 1·44 mM; [NaBr] = 0·20 M. Rotation speeds for each trace given in Hz.

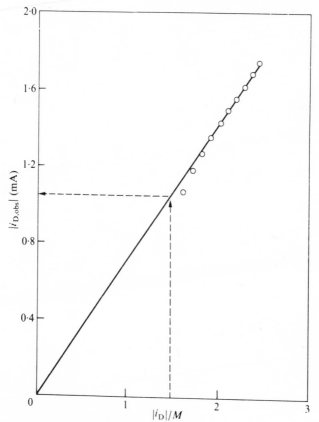

FIG. 8.4. Plot of observed disc current, $|i_{D,obs}|$, against calculated $|i_D|/M$. [allyl alcohol] = 1·44 mM; [NaBr] = 0·20 M; W = 5 Hz.

the first part of the curve are lower than those predicted for an infinitely fast reaction. This is because there is a significant contribution to the ring current from the bromine which penetrates across the reaction zone, and this means that the ring current rises to a particular level at a lower value of the disc current which is generating the bromine. In order to keep this deviation as small as possible it is important that the titration curve should be determined at a low rotation speed.

Fig. 8.4 shows how, knowing $\{1 - F(\alpha)\}$, we find $i_{D,K}$ directly; M is found from the gradient. By definition from eqn (7.5),

$$M = 12·27 \, r_1^2 n F D^{2/3} W^{1/2} \nu^{-1/6} c_\infty.$$

It is proportional to the concentration, and therefore we can very conveniently use the titration curve to measure the bulk concentration of C in any particular experiment. To do this we need to know D, but this

E

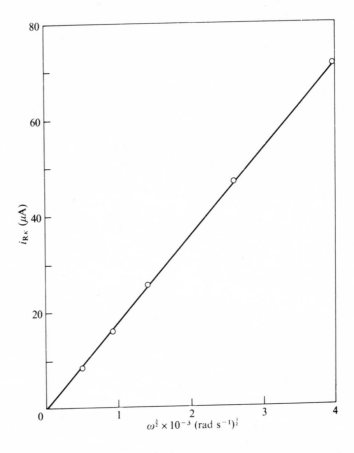

FIG. 8.5. Dependence of kinetic ring current, $i_{R,K}$, on $\omega^{3/2}$, [allyl alcohol] $=$ 1·44 mM; [NaBr] = 0·20 M.

has only to be determined once for any particular system, as described in the previous chapter.

Having found $i_{D,K}$, we return to the amplified traces in Fig. 8.3 and determine $i_{R,K}$ as shown by the dotted lines. N_K' is the ratio of these two quantities. Equation (8.7) predicts that $i_{R,K}$ should be proportional to $\omega^{3/2}$, and Fig. 8.5 shows results for the bromine/allyl alcohol system.

Normally, however, it is better to analyse the results with eqn (8.8), describing N_K'. The reason for this is that, as discussed at the start of the chapter, N_K' is a function only of the electrode geometry and κ. From eqn (8.1) experiments on the same system, with equal

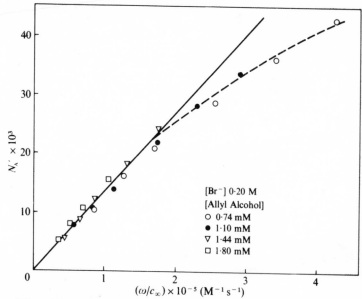

FIG. 8.6. Dependence of the kinetic collection efficiency, N_K', on (ω/c_∞)

values of $c_\infty^{1/2}/C^{1/3}$ or equal values of c_∞/ω, have the same values of κ. Thus values of N_K' from different experiments for the same system should lie on a common curve when plotted as a function of c_∞/ω. Equation (8.8) predicts that N_K' varies linearly with ω/c_∞, which is equivalent to κ^{-2}. Fig. 8.6 shows a plot of N_K' against ω/c_∞ for a set of experiments, at a particular $[Br^-]$, for the Br_2/allyl alcohol system. A straight line, as predicted by eqn (8.8), is obtained at low values of ω/c_∞, (high values of κ). However, this equation can no longer hold when Q' is large, so that either of the conditions (8.4) and (8.6) are broken. Q' is described by eqn (8.5):

$$ Q' = 0.06 \left(\frac{\omega}{k_2 c_\infty} \right)^{3/5}. $$

Thus, as ω/c_∞ increases, N_K' deviates below the predicted line. Indeed, in the limit, as $\kappa \to 0$ or $\omega/c_\infty \to \infty$, N_K' must tend to the constant value of N_0, the usual collection efficiency.

The advantage of using N_K' and eqn (8.8), rather than $i_{R,K}$ and eqn (8.7), is that all the experimental points lie on a common curve and it is easier to see where the necessary conditions for Q' hold. The rate constant is obtained from the gradient as ω/c_∞ tends to zero. Q' also depends upon k_2 and the slower the reaction the more likely it is that the critical conditions will not hold.

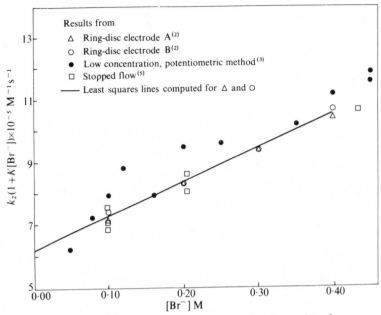

FIG. 8.7. Dependence of $k_2(1 + K[Br^-])$ on $[Br^-]$.

Both Br_2 and Br_3^- brominate allyl alcohol and the observed rate constant in any particular $[Br^-]$ is given by

$$k_2(1 + K[Br^-]) = k_{Br_2} + k_{Br_3^-} K[Br^-] \qquad (8.9)$$

where

$$K = \frac{[Br_3^-]}{[Br_2] \times [Br^-]} \simeq 16.[4]$$

Results for the left-hand side of this equation, obtained by the ring-disc technique, for two electrodes of different geometries are compared with Bell and Atkinson's[3] results in Fig. 8.7. Also shown are some more recent stopped flow determinations carried out by Dr. D.J. Barnes in Professor Bell's laboratory at Stirling.[5] Good agreement between the different techniques is found.

Numerical Calculations

The numerical technique described in Chapter 2 has been used by Bard and Prater[6] to calculate N_K' as a function of κ. The result of their numerical calculations for one of our electrodes is shown in Fig. 8.8. Agreement to about 20 per·cent is obtained for the high values of κ. The numerical calculations show clearly the curvature in N_K' as it tends towards the limit of N_0 for $\kappa \to 0$.

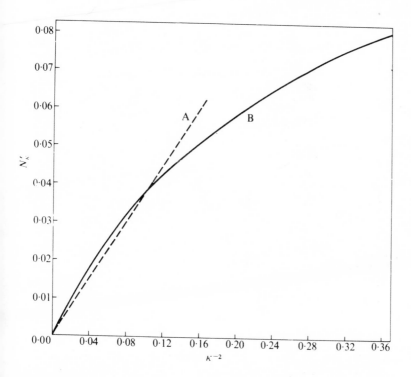

FIG. 8.8. N'_K as a function of κ^{-2} – comparison of analytical and numerical solutions. A, Analytical solution – eqn (8.8). B, Numerical solution.[5] Electrode geometry: $r_1 = 0.4770$ cm; $r_2 = 0.4869$ cm; $r_3 = 0.5222$ cm.

The method of calculating eqns (8.7) and (8.8) depended on the method of moments and therefore may only be accurate to about 20 per cent. However, the experimental results for the bromination of allyl alcohol agree better with the analytical solution than with the numerical computation. So at the moment there is some doubt about the exact value of the numerical constants in eqns (8.7) and (8.8). If we believe the results for allyl alcohol and the analytical theory then the values given are correct. If we believe the numerical results then the values should be multiplied by 1.25. More work on reactions of known rate constant is needed to settle this point.

The Reaction between Br_2 and As (III)

We have used the ring-disc technique to measure the rate constant for the oxidation of As (III):[7]

$$\text{disc} \qquad Br^- \longrightarrow Br_2 ;$$

$$\text{solution} \qquad Br_2 + As(III) \xrightarrow{k_2} As(V) + 2\,Br^- ;$$

$$\text{ring} \qquad Br_2 \longrightarrow Br^- .$$

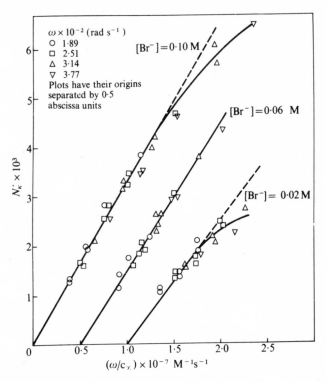

FIG. 8.9. Dependence of N_K' on (ω/c_∞) for the $Br_2/As(III)$ system.

This is a much faster reaction than the bromination of allyl alcohol with values of k_2 of about $10^8 M^{-1} s^{-1}$. Fig. 8.9 shows some plots of N_K' against ω/c_∞, where c_∞ is the bulk concentration of As(III). The reaction was carried out in $1\,M\,H_2SO_4$, and at constant ionic strength. The rate constants calculated from eqn (8.8) are analysed in the same way as eqn (8.9) and plotted in Fig. 8.10 to show the effect of $[Br^-]$. The rate constants for oxidation by Br_2 and by Br_3^- are

$$k_{Br_2} = (1.13 \pm 0.03) \times 10^8 M^{-1} s^{-1},$$

$$k_{Br_3^-} = (4.65 \pm 0.28) \times 10^7 M^{-1} s^{-1}.$$

Similar studies have been carried out by Bruckenstein and Johnson on the oxidation of arsenic by iodine.[8]

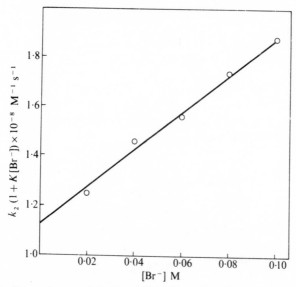

FIG. 8.10. Dependence of $k_2(1 + K[Br^-])$ on $[Br^-]$ for the $Br_2/As(III)$ system.

Thus the ring-disc electrode is a useful technique for studying the kinetics of fast irreversible reactions in solution. We estimate that rate constants up to $\sim 10^9 \, M^{-1} s^{-1}$ can be measured in this way. This is with usual rotation speeds of $\sim 50 \, Hz$. Since $i_{R,K}$ depends on $\omega^{3/2}$ it is worth increasing ω for this particular application. There are not many other techniques that are capable of measuring such fast, irreversible reactions.

References

1. ALBERY, W.J. and BRUCKENSTEIN, S. (1966) *Trans. Faraday Soc.* 62, 2584.

2. ALBERY, W.J., HITCHMAN, M.L., and ULSTRUP, J. (1969) *Trans. Faraday Soc.* 65, 1101.

3. BELL, R.P. and ATKINSON, J.R. (1963) *J. chem. Soc.* 3260.

4. SCAIFE, D.B. and TYRRELL, H.J.V. (1958) *J. chem. Soc.* 386.

5. BARNES, D.J. (1969) Private communication.

6. BARD, A.J. and PRATER, K.B. (1970) *J. electrochem. Soc.* 117,335.

7. ALBERY, W.J. and HITCHMAN, M.L. (1968) Unpublished results.

8. BRUCKENSTEIN, S. and JOHNSON, D.C. (1968) *J. Am. chem. Soc.* 90, 6592.

9

HOMOGENEOUS KINETICS – FIRST ORDER

Introduction

WE now consider the case where the electrogenerated intermediate decomposes on its way from the disc to the ring electrode by first-order kinetics. This may be a genuine unimolecular reaction, or it may be reacting with the solvent, or any other species, the concentrati of which is much larger than that of the intermediate. The system is

$$\text{disc} \qquad A \longrightarrow B\,;$$

$$\text{solution} \qquad B \overset{k_1}{\longrightarrow} \text{products}\,;$$

$$\text{ring} \qquad B \longrightarrow A\,.$$

Therefore in this case, unlike the second-order case, we have only to describe the concentration profile for B. If B is reacting with some other species C in much larger concentration, then

$$k_1 = k_2 c_\infty. \tag{9.1}$$

Mathematical Theory

The differential equation for B is the same as eqn (3.6) except that we add the extra kinetic term

$$D\frac{\partial^2 b}{\partial z^2} - v_r\frac{\partial b}{\partial r} - v_z\frac{\partial b}{\partial z} - k_1 b = 0. \tag{9.2}$$

This equation can be solved analytically for thin-gap thin-ring electrodes, that is electrodes with low values of α and β. The reason for this restriction is that in order to solve the equation in the zones of the gap and the ring we have to neglect the term, $v_z\,\partial b/\partial z$, describi normal convection. This is justified for thin-gap thin-ring electrodes because the concentration profile in the zones of the gap and the ring for such an electrode is determined by the surface concentration, and the concentration gradient at $z = 0$ on the outside edge of the disc electrode; the radial distances are so small that it is not affected in these zones by the normal convection on the outside edge of the disc electrode.

We can solve eqn (9.2) with $k_1 = 0$ and neglect the normal convection to give an approximate collection efficiency N_0'' where

$$N_0'' = -\tfrac{2}{3}\alpha' - \tfrac{2}{3}\beta' + \tfrac{2}{3}(\alpha' + \beta')F(\alpha'/\beta') +$$
$$+ (\beta')^{2/3} - 3^{1/2}\pi^{-1}(\beta')^{2/3}(\alpha')^{1/3},$$
$$\alpha' = 3\ln(r_2/r_1),$$

and

$$\beta' = 3\ln(r_2/r_1).$$

It can be shown that for thin-gap thin-ring electrodes

$$\alpha = (r_2/r_1)^3 - 1 \longrightarrow \alpha'.$$
$$\beta = (r_3/r_1)^3 - (r_2/r_1)^3 \longrightarrow \beta'.$$

and

$$N_0 \longrightarrow N_0''.$$

Indeed the difference between N_0 and N_0'' is a good measure of how justified is this approximation.

Equation (9.2) is transformed with the same substitutions as in eqn (3.7) to give

$$\frac{\partial^2 u}{\partial w^2} - rw\frac{\partial u}{\partial r} + w^2\frac{\partial u}{\partial w} - \kappa^2 u = 0, \tag{9.3}$$

where

$$\kappa = (k_1/D)^{1/2}(D/C)^{1/3}, \tag{9.4}$$

and this equation obeys the same boundary conditions.

In the zone of the disc $\partial u/\partial r = 0$ and the resulting equation has been solved numerically by Hale.[1] He has shown that a good analytical approximation for u_*, the value of u on the surface of the electrode, is

$$u_* = u_0 \underset{r<r_1}{} = \kappa^{-1}\tanh(A_1\kappa). \tag{9.5}$$

This has the correct limiting properties in that for small κ, $u_* = A_1$, as in Chapter 2; for large κ, $u_* = \kappa^{-1}$, since for fast kinetics $u_* = \kappa^{-1}e^{-\kappa w}$. Equation (9.5) holds to within 3 per cent for all values of κ; however, it does not describe accurately the approach of u_* to u_0 for small values of κ. Since this difference is important for small values of κ it is better to use Hale's numerical table of $v(\lambda)$, where $\lambda = (A_1\kappa)^2$, to obtain a series approximation:

$$u_* = A_1(1 - 0{\cdot}372\,\lambda + 0{\cdot}146\,\lambda^2 \ldots).\tag{9.6}$$

In the zones of the ring and the gap we define a new concentration variable u', where

$$u' = \kappa\,(u_{r_1} - u)$$

and u_{r_1} is the value of u at $r = r_1$. Neglecting the normal convection term and transforming the variables with

$$y = \kappa w, \qquad \xi_{n,\kappa} = \kappa^3 \ln(r/r_n),$$

we obtain

$$y\,\frac{\partial u'}{\partial \xi_{n,\kappa}} = \frac{\partial^2 u'}{\partial y^2} - u',$$

where $n = 1$ for the zone of the gap, and $n = 2$ for the zone of the ring.
    The boundary conditions for u' are

$$y \longrightarrow \infty, \qquad u' \longrightarrow 0;$$

disc $r < r_1,$ $y = 0,$ $\partial u'/\partial y = -1;$

gap $r_1 < r < r_2,$ $y = 0,$ $\partial u'/\partial y = 0;$

ring $r_2 < r < r_3,$ $y = 0,$ $u' = \kappa u_*.$

We then obtain[2]

$$N_\kappa'' = \frac{-2}{\kappa^3}\left\{ \xi_{2,\kappa}' + \mathcal{L}_2^{-1}\mathcal{L}_1^{-1}\,\frac{1}{s_2}\left(\frac{\partial \bar{\bar{u}}'}{\partial y}\right)_0 \begin{array}{l} {\scriptstyle \xi_{1,\kappa} = \xi_{1,\kappa}'} \\ {\scriptstyle \xi_{2,\kappa} = \xi_{2,\kappa}'} \end{array} \right\}$$

where

$$\xi_{n,\kappa}' = \kappa^3 \ln(r_{n+1}/r_n),$$

s_n is the transformed variable of ξ_n and

$$\frac{1}{s_2}\left(\frac{\partial \bar{\bar{u}}'}{\partial y}\right)_{y=0} = \frac{Ai'(s_2^{-2/3})\,\kappa u_*}{s_1 s_2^{5/3}\,Ai(s_2^{-2/3})} +$$

$$+\ \frac{s_1^{1/3}\,Ai'(s_1^{-2/3})\,Ai(s_2^{-2/3}) - s_2^{1/3}\,Ai(s_1^{-2/3})\,Ai'(s_2^{-2/3})}{s_1^{4/3}\,s_2(s_1 - s_2)\,Ai'(s_1^{-2/3})\,Ai(s_2^{-2/3})}.$$

$$\tag{9.7}$$

An analytical inversion of this function has not proved possible. However, a computer programme in which the Airy functions are expanded and the polynomial inverted term by term has been developed and work well. The calculation has to be performed once for any electrode to give values of N_κ'' as a function of κ and the geometry of the electrode
    In general, eqn (9.5) is used for u_*, except for small values of κ when eqn (9.6) is to be preferred.

For κ smaller than 1 an approximate analytical expression may be obtained[2] using a much smaller number of terms:

$$N_\kappa'' = N_0'' - (\beta')^{2/3} (1 - A_1^{-1} u_*) +$$
$$+ \tfrac{1}{2} A_1^{-1} A_2^2 \kappa^2 u_* (\beta')^{4/3} - 2 A_2 \kappa^2 T_2. \qquad (9.8)$$

The small T_2 term is given in Appendix 4. This expression does not differ from the full computer calculation for $\kappa \leqslant 1$.

These equations hold for a thin-gap thin-ring electrode, where $N_0 \simeq N_0''$. In many cases it is difficult to construct an electrode that fulfils this condition exactly. In practice there may be a 20 per cent difference between N_0 and N_0''. We therefore include a correction for this difference, by writing

$$N_\kappa = \left[1 + \{1 - \tanh^2(A_1 \kappa)\}\left\{\frac{N_0}{N_0''} - 1\right\} \right] N_\kappa''. \qquad (9.9)$$

The reason for choosing $\tanh(A_1 \kappa)$ is that in eqn (9.5) for u_* it describes the balance between kinetics and normal convection. As κ increases the terms describing normal convection become less and less important since the intermediate is all destroyed before it reaches the outside of the diffusion layer. The square bracket tends to 1 when κ is large since no correction is then necessary. It tends to N_0/N_0'' when κ is small, and therefore N_κ obeys the correct limiting condition

$$N_\kappa \longrightarrow N_0 \quad \text{when} \quad \kappa \longrightarrow 0.$$

Fig. 9.1 shows a plot of N_κ against κ for two electrodes. Two sets of figures are calculated, depending on which of eqns (9.5) and (9.6) we use for u_*. As can be seen, for $\kappa \sim 0\cdot3$ the same values for N_κ are obtained from either equation. We therefore use eqn (9.5) for $\kappa > 0\cdot3$, and eqn (9.6) for $\kappa < 0\cdot3$.

For $\kappa > 1$, since it may be difficult for most people to invert eqn (9.7) either analytically or numerically, we have fitted the output of our computer programme, which does the inversion numerically, to a simple analytical expression:

$$\ln N_\kappa - \ln N_0 = -a\kappa + b - \ln\{1 + 1/(\kappa + c)\}. \qquad (9.10)$$

This expression contains three parameters a, b, and c. We have calculated values of these parameters for electrodes of common radius ratios, and the results are tabulated in Appendix 5, together with a figure showing the percentage deviations. Generally the equation fits to better than 2 per cent for values of κ in the range

$$1\cdot0 < \kappa < 4\cdot0.$$

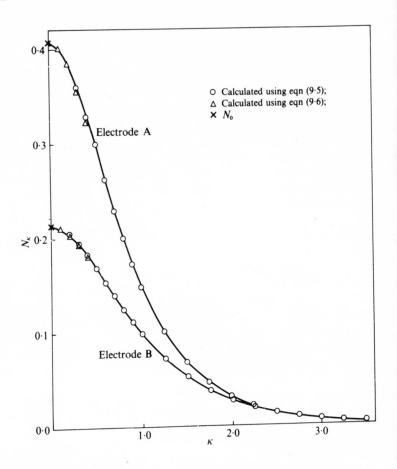

FIG. 9.1. Calculated N_κ values plotted against κ for two electrodes. Electrode dimensions

	r_1 (cm)	r_2 (cm)	r_3 (cm)	N_0	N_0''
A	0·3675	0·3777	0·4835	0·407	0·286
B	0·4770	0·4869	0·5222	0·213	0·189

Numerical Calculations

Bard and Prater[4] have applied their numerical technique to the calculation of N_κ as a function of κ for one of our electrodes. Table 9.1 compares the results from the two methods of calculation. Very good agreement is found. Also included are results from the analytica eqn (9.8) for $\kappa < 1$.

TABLE 9.1

Comparison of values of N_K obtained numerically,[4,5] by computer inversions of eqn (9.7)[3] and from the analytical expression, eqn (9.8)[2]

Electrode dimensions $r_1 = 0.4770$ cm; $r_2 = 0.4869$ cm; $r_3 = 0.5220$ cm.

	Kinetic collection efficiency N_K		
K	Numerical solution [5]	Computer inversion of eqn (9.7)[†]	Analytical expression, eqn (9.8)[†]
0.100	0.209	0.210	-
0.224	0.201	0.201	-
0.316	0.192	0.193	0.192
0.447	0.176	0.176	0.174
0.548	0.162	0.162	0.159
0.633	0.150	0.149	0.147
0.707	0.139	0.138	0.136
0.775	0.129	0.128	0.125
0.857	0.120	0.119	0.114
0.894	0.112	0.112	0.107
0.949	0.105	0.105	0.100
1.000	0.098_6	0.098_4	0.094
1.225	0.073_8	0.074_2	-
1.414	0.057_1	0.058_1	-
1.732	0.036_6	0.038_0	-
2.000	0.024_8	0.026_4	-
2.236	0.017_4	0.019_0	-

[†] N_K obtained from N_K'' using eqn (9.9).

Experimental Results

Three reactions of known rate constant have been used to test the theory of first-order kinetics.[6] The systems are:

System I

$$\text{disc} \qquad Br^- \longrightarrow \tfrac{1}{2} Br + e;$$

$$\text{solution} \qquad Br_2 + CH_3OC_6H_5 \longrightarrow HBr + CH_3OC_6H_4Br;$$

$$\text{ring} \qquad e + \tfrac{1}{2} Br_2 \longrightarrow Br^-.$$

System II

$$\text{disc} \qquad Br^- \longrightarrow \tfrac{1}{2} Br_2 + e;$$

$$\text{solution} \qquad Br_2 + m\,CH_3OC_6H_4F \longrightarrow HBr + CH_3OC_6H_3FBr;$$

$$\text{ring} \qquad e + \tfrac{1}{2} Br_2 \longrightarrow Br^-.$$

System III

disc	$Fe^{3+} + e \longrightarrow Fe^{2+};$
solution	$VO_2^+ + Fe^{2+} + 2H^+ \longrightarrow VO^{2+} + Fe^{3+} + H_2O;$
ring	$Fe^{2+} \longrightarrow Fe^{3+} + e.$

A typical series of ring-current traces are shown for System I in Fig. 9.2. The ring current increases linearly with the disc current, the

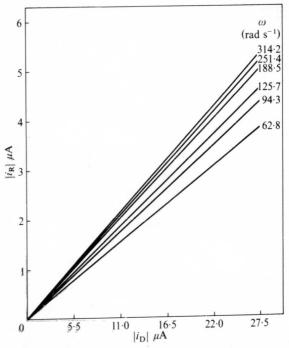

FIG. 9.2. Experimental $i_R - i_D$ kinetic traces. Electrode B; System I; $[Br^-] = 0.378\,M$; [anisole] $= 0.579\,mM$.

gradient being higher the higher the rotation speed. The gradient is, of course, equal to N_K:

$$|i_R| = N_K |i_D|.$$

To convert a value of N_K measured from the i_R/i_D trace we use the theoretical plot shown in Fig. 9.1 to obtain the corresponding value of κ.

From eqn (9.4), and remembering that

$$k_1 = k_2 c_\infty, \tag{9.1}$$

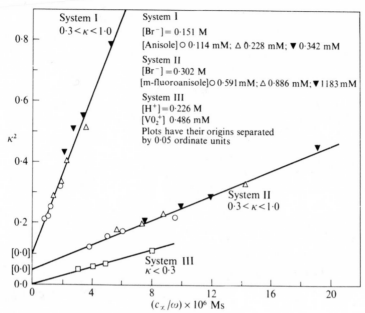

FIG. 9.3. Dependence of κ^2 on c_∞/ω.

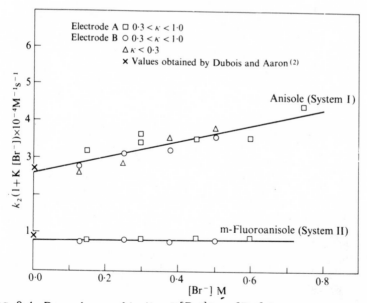

FIG. 9.4. Dependence of $k_2(1 + K[Br^-])$ on $[Br^-]$ for Systems I and II.

$$\kappa^2 = 1\cdot56 \left(\frac{\nu}{D}\right)^{1/3} \cdot \frac{c_\infty}{\omega} \cdot k_2 \quad (\omega \text{ in rad s}^{-1})$$

$$= 0\cdot248 \left(\frac{\nu}{D}\right)^{1/3} \cdot \frac{c_\infty}{W} \cdot k_2 \quad (W \text{ in Hz}).$$

Fig. 9.3 shows typical plots for the three systems of κ^2 against c_∞/ω. In agreement with theory good straight lines are obtained. From the gradient we can calculate k_2 as long as we know D and ν. The measurement of ν with an Ostwald vicometer is simple; D can be measured directly for the limiting current at a rotating-disc electrode, or else by a new technique discussed below. In any case both quantities only appear to the power of $\frac{1}{3}$.

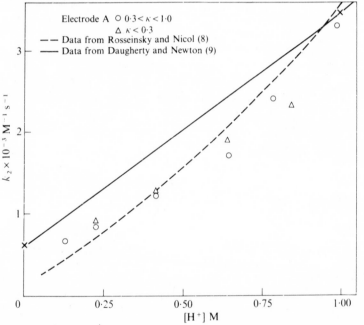

FIG. 9.5. Dependence of observed rate constant upon $[H^+]$ for System III.

Values for the rate constant for the bromination of the anisoles are shown in Fig. 9.4, together with the points obtained by Dubois and Aaron.[7] The factor $(1 + K[Br^-])$ allows for the fact that some of the Br_2 generated on the disc electrode is present as Br_3^- in the solution. Results for System III are shown in Fig. 9.5.

Diffusion Coefficients

We have recently developed a simple new technique for measuring diffusion coefficients using a rotating disc electrode. The decay of the limiting current at the disc electrode is measured as a function of time as the bulk solution is depleted of the species that is being destroyed on the electrode.

Then we may write from eqn (2.16):

$$V \frac{\partial c_\infty}{\partial t} = - \frac{i_L}{nF} = - \frac{1 \cdot 554 \, \pi \, r_1^2 D^{2/3} W^{1/2}}{\nu^{1/6}} c_\infty,$$

where V is volume of the solution.

Thus the bulk concentration and the limiting current, which is proportional to it, decay in a first-order fashion with a rate constant of

$$\frac{1 \cdot 554 \, \pi \, r_1^2 D^{2/3} W^{1/2}}{V \, \nu^{1/6}} .$$

A plot of log i_L against t gives this quantity from which D can be calculated. In order that the experiment should not take too long we have reduced V from our normal operating volume of $\sim 100 \, cm^3$ to $10 \, cm^3$. We have also used a larger electrode ($r_1 \sim 0 \cdot 3 \, cm$). Under these conditions and $W = 20 \, Hz$, the half-time is $\sim 60 \, min$.

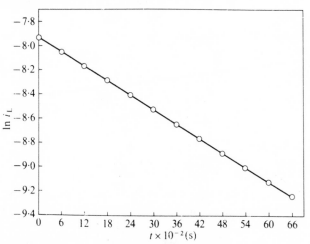

FIG. 9.6. Plot of ln i_L against time. $W = 20 \cdot 0 \, Hz$; $r_1 = 0 \cdot 3611 \, cm$; $V = 10 \cdot 50 \, cm^3$; $\nu = 8 \cdot 525 \times 10^{-3} \, cm^2 \, s^{-1}$.

Some typical results are shown in Fig. 9.6 for D for $Fe(CN)_6^{4-}$ in $1 \, M \, KCl$.[10] The value of $(0 \cdot 633 \pm 0 \cdot 002) \times 10^{-5} \, cm^2 \, s^{-1}$ is in very good agreement with that of von Stackleberg, Pilgram, and Toome[11] – $(0 \cdot 632 \pm 0 \cdot 003) \times 10^{-5} \, cm^2 \, s^{-1}$. The advantage of the method is that we do not need to know either c_∞ or n to determine D. The normal use of the limiting current requires both these quantities. Indeed, if c_∞

is known an experiment of this type gives an unambiguous measurement of D and of n.

The Range of Measurable Rate Constants

The range of rate constants that can be measured by this first-order technique is a very wide one. Values of κ between $0\cdot 2$ and 5 may be measured. For $\kappa < 0\cdot 2$ the difference between N_K and N_0 is too small, and for $\kappa > 5$ so little intermediate arrives at the ring electrode that the current is too small to be measured accurately.

Taking values of D of $10^{-5} \text{cm}^2 \text{s}^{-1}$ and ν of $10^{-2} \text{cm}^2 \text{s}^{-1}$, we have the following condition:

$$1\cdot 6 \times 10^{-2} < \frac{k_1}{W} < 10.$$

W may be varied easily between 2 and $100\,\text{Hz}$ so that we obtain

$$3 \times 10^{-2}\,\text{s}^{-1} < k_1 < 10^3\,\text{s}^{-1},$$

or the half-life of the intermediate lies in the following range:

$$20\,\text{s} > t_{1/2} > 7 \times 10^{-4}\,\text{s}.$$

If the intermediate is reacting with a species the concentration of which can be varied so that

$$k_1 = k_2\, c_\infty \tag{9.1}$$

then c_∞ should lie in the following range:

$$1\,\text{M} > c_\infty > 10^{-4}\,\text{M},$$

so that we have for k_2

$$3 \times 10^{-2}\,\text{M}^{-1}\text{s}^{-1} < k_2 < 10^7\,\text{M}^{-1}\text{S}^{-1}.$$

The largest second-order rate constant that can be measured by this technique is therefore about two orders of magnitude less than the second-order method. This is because, in the second-order method the intermediate travels most of the way to the ring electrode in its own friendly environment before reaching the reaction front where it encounters the hostile reagent that destroys it. In the first-order method the whole of the solution between the disc and the ring is occupied by the hostile reagent and the intermediate suffers attrition during the whole of its passage; in this type of guerrilla warfare there is no well defined front.

To survive in detectable quantities the intermediate has to have a lifetime of at least $1\,\text{ms}$. Thus intermediates that are destroyed by the solvent more rapidly than this cannot be detected. If the technology of the construction of ring-disc electrodes can be improved so that electrodes with thinner gaps can be made then it should be possible to detect shorter-lived intermediates generated in electrode mechanisms.

The ring-disc electrode is therefore a powerful technique for the study of homogeneous kinetics. A wide range of rate constants can be measured to an accuracy of a few per cent. The restrictions in the use of the technique concerns the electrochemistry. These are:

(1) One of the reagents must be generated on the disc electrode.

(2) Other species in the solution must not interfere with this generation. This restriction does not apply to the species C in the second-order technique since for a fast reaction its concentration is zero on the disc electrode.

(3) All the intermediate that reaches the ring electrode must be destroyed.

(4) Other species in the solution, including the products of the homogeneous reaction must not interfere with this detection at the ring electrode.

One advantage of solid electrodes is that we can choose the electrode material so that these conditions are best met.

References

1.   HALE, J.M. (1964) *J. electroanalyt. Chem,* 8, 332.

2.   ALBERY, W.J. (1967) *Trans. Faraday Soc.* 63, 1771.

3.   ALBERY, W.J. (1970) *Trans. Faraday Soc.* In press.

4.   BARD, A.J. and PRATER, K.B. (1970) *J. electrochem. Soc.* 117, 335.

5.   BARD, A.J. and PRATER, K.B. (1969) Private communication.

6.   ALBERY, W.J., HITCHMAN, M.L., and ULSTRUP, J. (1968) *Trans. Faraday Soc.* 64, 2831.

7.   DUBOIS, J.F. and AARON, J.J. (1964) *C.r. hebd. Séanc. Acad. Sci., Paris* 258, 2313.

8.   DOUGHERTY, N.A. and NEWTON, T.W. (1963) *J. phys. Chem. Ithaca* 67, 1090.

9.   ROSSEINSKY, D.R. and NICOL, M.J. (1966) *Electrochim. Acta* 11, 1069.

10. ALBERY, W.J. and HITCHMAN, M.L. (1970) Unpublished results.

11. STACKLEBERG, M. VON, PILGRAM, M., and TOOME, V. (1953) *Z. Elektrochem.* 57, 342.

10

TRANSIENT CURRENTS

Introduction

ALL the applications of the ring-disc electrode discussed hitherto have been confined to steady-state conditions. The concentration profiles and currents have not been functions of time. However, just as techniques like chronopotentiometry are useful in single-electrode systems, so the transient behaviour of the ring and disc currents yields information that could not be obtained by steady-state studies. For the generation of the intermediate, B, we now restore the $\partial b/\partial t$ term to eqn (2.4) to give

$$\frac{\partial b}{\partial t} = D\frac{\partial^2 b}{\partial z^2} - v_z \frac{\partial b}{\partial z} - v_r \frac{\partial b}{\partial r}.$$

We now define the non-dimensional time variable τ, where

$$\tau = t\,C^{2/3}D^{1/3}, \tag{10.1}$$

and we normalize the concentration with

$$b - b_{t=0} = -\psi \left\{ \left(\frac{\partial b}{\partial w}\right)_{\substack{w=0 \\ r<r_1 \\ t\to\infty}} - \left(\frac{\partial b}{\partial w}\right)_{\substack{w=0 \\ r<r_1 \\ t=0}} \right\},$$

where w has the usual definition

$$w = C^{1/3}D^{-1/3}\,z.$$

The equation then becomes

$$\frac{\partial \psi}{\partial \tau} = \frac{\partial^2 \psi}{\partial w^2} - r\,w\,\frac{\partial \psi}{\partial w} + w^2\,\frac{\partial \psi}{\partial w}, \tag{10.2}$$

with boundary conditions

$$\tau = 0, \quad \psi = 0;$$

$$w \to \infty, \quad \psi \to 0;$$

134

disc electrode $r < r_1,$ $w = 0,$ $\dfrac{\partial \psi}{\partial w} = f(\tau);$

gap $r_1 < r < r_2,$ $w = 0,$ $\dfrac{\partial \psi}{\partial w} = 0;$

ring electrode $r_2 < r < r_3,$ $w = 0,$ $\psi = 0.$

The function $f(\tau)$ describes the variation of the flux of the intermediate
at the surface of the disc electrode:

$$f(\tau) = - \frac{\left(\dfrac{\partial b}{\partial z}\right)_{\substack{z=0 \\ r<r_1 \\ \tau}} - \left(\dfrac{\partial b}{\partial z}\right)_{\substack{z=0 \\ r<r_1 \\ \tau=0}}}{\left(\dfrac{\partial b}{\partial z}\right)_{\substack{z=0 \\ r<r_1 \\ \tau\to\infty}} - \left(\dfrac{\partial b}{\partial z}\right)_{\substack{z=0 \\ r<r_1 \\ \tau=0}}}. \tag{10.3}$$

The function is designed so that it varies from 0 at $t = 0$ to -1 as
$t \to \infty$.

We define a function N_t to describe the ring current:

$$N_t = -\frac{n_D(i_{R,t} - i_{R,t=0})}{n_R(i_{D,t\to\infty} - i_{D,t=0})}. \tag{10.4}$$

The minus sign allows for the fact that the ring and disc currents are
in opposite directions, and n_D and n_R describe the number of electrons
in the overall reaction at the disc and ring electrodes. N_t varies between
0 at $t = 0$ and N_0, the steady-state collection efficiency, at $t = \infty$.

In terms of the variable ψ,

$$N_T = \frac{2}{r_1^2} \int_{r_2}^{r_3} \left(\frac{\partial \psi}{\partial w}\right)_{w=0} r\, dr.$$

We now transform eqn (10.2) with respect to time, and we write the
variable as σ so that it will not be confused with s which has been
used for distance variables:

$$\sigma \bar{\psi} = \frac{\partial^2 \bar{\psi}}{\partial w^2} - r\, w \frac{\partial \bar{\psi}}{\partial r} + w^2 \frac{\partial \bar{\psi}}{\partial w}.$$

This has the same form as the differential equation (9.3) describing
decomposition by first-order kinetics:

$$\kappa^2 u = \frac{\partial^2 u}{\partial w^2} - r\, w \frac{\partial u}{\partial r} + w^2 \frac{\partial u}{\partial w}. \tag{9.3}$$

The boundary conditions are the same except that on the disc electrode at $w = 0$ $\partial \bar{\psi}/\partial w = \bar{f}(\sigma)$, whereas $\partial u/\partial w = -1$. So we now define

$$\bar{\psi}' = \frac{\bar{\psi}}{-\bar{f}(\sigma)} .$$

Then $\bar{\psi}'$ obeys the differential equation (9.3) with boundary conditions:

$$w \to \infty, \quad \psi' \to 0;$$

disc electrode $\qquad\qquad r < r_1, \quad w = 0, \quad \frac{\partial \bar{\psi}'}{\partial w} = -1;$

gap $\qquad\qquad\qquad r_1 < r < r_2, \quad w = 0, \quad \frac{\partial \bar{\psi}'}{\partial w} = 0;$

ring electrode $\qquad\quad r_2 < r < r_3, \quad w = 0, \quad \bar{\psi}' = 0;$

and

$$N_T = -\frac{2\bar{f}(\sigma)}{r_1^2} \int_{r_2}^{r_3} \left(\frac{\partial \bar{\psi}'}{\partial w}\right)_{w=0} r \, dr.$$

But

$$N_K = \frac{2}{r_1^2} \int_{r_2}^{r_3} \left(\frac{\partial u}{\partial w}\right)_{w=0} r \, dr.$$

Therefore

$$\bar{N}_T = -N_{\sigma^{1/2}} \, \bar{f}(\sigma), \tag{10.5}$$

where

$$N_{\sigma^{1/2}} = N_K,$$

when N_K is calculated for a value of κ corresponding to $\sigma^{1/2}$. The calculation of N_K as a function of κ is described in Chapter 9.

Ring Transient for Galvanostatic Step at the Disc

The analytical inversion of N_T back into real time is in general impossible since $N_{\sigma^{1/2}}$ and $\bar{f}(\sigma)$ are themselves complicated functions of σ. However, the inversion can be achieved for the case of a step in the current at the disc and for $\tau \sim$ or > 1.

The step in the current at the disc gives the particularly simple form for $\bar{f}(\sigma)$:

$$\bar{f}(\sigma) = -\frac{1}{\sigma}. \tag{10.6}$$

For $\sigma^{1/2} < 1$, corresponding to $\tau \sim$ or > 1 and $\kappa < 1$, we can write

$$N_{\sigma 1/2} \simeq \frac{N_0}{N_0''} N_{\sigma 1/2}''$$

$$\simeq N_0 - \frac{N_0}{N_0''} [(\beta')^{2/3}(1 - A_1^{-1}\bar{\psi}_*') + \tfrac{1}{2} A_1^{-1} A_2^2 \sigma \bar{\psi}_*'(\beta')^{4/3}],$$

Where the small T_2 term has been omitted, and from eqns (9.5) and (9.6),

$$\bar{\psi}_* = \frac{\tanh(A_1 \sigma^{1/2})}{\sigma^{1/2}}, \quad \text{for} \quad \sigma > 0\cdot 1,$$

and

$$\bar{\psi}_* = A_1(1 - 0\cdot 372\,\lambda + 0\cdot 146\,\lambda^2 \ldots),$$

where

$$\lambda = A_1^2\,\sigma, \quad \text{for} \quad \sigma < 0\cdot 1.$$

On substituting and carrying out the inversion we obtain two equations.[1] First, for $\sigma > 0\cdot 1$,

$$N_T = N_0 - \frac{N_0}{N_0''} [(\beta')^{2/3}\{1 - M(\tfrac{1}{4}\pi^2\rho)\} - \tfrac{1}{2}A_2^2 A_1^{-2}(\beta')^{4/3} L(\rho)], \tag{10.7}$$

where

$$\rho = A_1^{-?}\,\tau,$$

$$M(\tfrac{1}{4}\pi^2\rho) = 1 - 8\pi^{-2} \sum_{n=1}^{\infty} (2n-1)^{-2} \exp\{-(2n-1)^2(\tfrac{1}{4}\pi^2\rho)\},$$

and

$$L(\rho) = 2 \sum_{n=1}^{\infty} \exp\{-(2n-1)^2(\tfrac{1}{4}\pi^2\rho)\}.$$

The function M is tabulated by McKay,[2] who first worked it out when he was employed by the Council of the British Boot Shoe and Allied Trades Research Association to study the diffusion of rainwater through leather soles. The function L is tabulated by Hale;[3] he worked it out while he was at Farnborough. What he was doing there is still an official secret.

Secondly, for $\sigma < 0\cdot 1$,

$$N_T = N_0 - \frac{N_0}{N_0''}\frac{B^2}{B'}\exp\left(-\frac{B\tau}{B'}\right), \tag{10.8}$$

where[1]

$$B = 0\cdot 617(\beta')^{2/3} - 0\cdot 266(\beta')^{4/3}$$

and

$$B' = 0\cdot 402(\beta')^{2/3} - 0\cdot 164(\beta')^{4/3};$$

B and B' are just functions of the electrode geometry.

These two equations have been tested for two systems:[4]

	System I	System II
disc	$Fe(CN)_6^{3-} \longrightarrow Fe(CN)_6^{4-}$	$Br^- \longrightarrow Br_2$
ring	$Fe(CN)_6^{4-} \longrightarrow Fe(CN)_6^{3-}$	$Br_2 \longrightarrow Br^-$
disc current step (mA)	$-0.85 \longleftrightarrow -1.56$	$0 \longleftrightarrow 0.21$

The disc current was stepped either up or down between the two values shown.

Fig. 10.1 shows two typical oscilloscope traces of the ring current for system II for a positive and a negative step in the disc current.

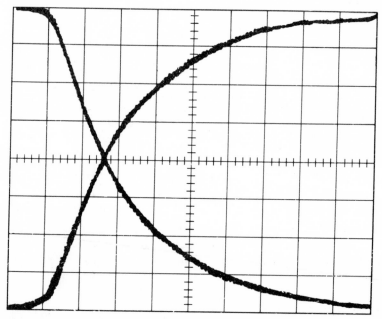

FIG. 10.1. Oscilloscope traces of typical ring current transient obtained for galvanostat step at the disc. x-axis, Time 1 division = 20 ms. y-axis, Ring current 1 division = 33·3 μA. System II: ω = 249·0 rad s^{-1}; Δi_D = 1·25 mA.

Note that as predicted by theory the transients are mirror images of each other. Curves such as these are analysed in the following manner. From eqn (10.7) and (10.8) a theoretical curve of N_τ/N_0 against τ is calculated. These curves are shown in Fig. 10.2. They are used to relate a value of N_t/N_0 at time t from the oscilloscope trace with the

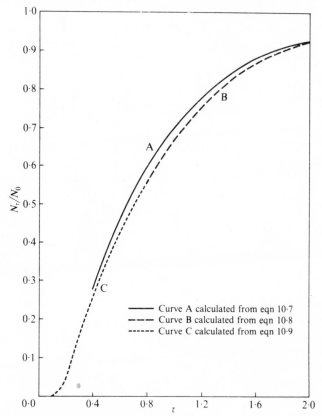

FIG. 10.2. Theoretical curves for N_T/N_0 as a function of τ.

corresponding value of τ. Each value of t is then plotted against its
value of τ. Typical plots are shown for the two systems in Figs. 10.3
and 10.4. The lines are the lines predicted by eqn (10.1):

$$\tau = t C^{2/3} D^{1/3}.$$

Complete data for both systems, for positive and negative steps, and
for different rotation speeds are given in Tables 10.1 and 10.2. Good
agreement between theory and experiment is found. It is satisfactory
that eqn (10.7) fits the data better at short times while eqn (10.8)
works better at long times.

Bard and Prater's numerical method[5] also can be used to calculate
ring current transients. Fig. 10.5 compares a numerically calculated
transient with results from eqns (10.7) and (10.8). Reasonable agreement is found.

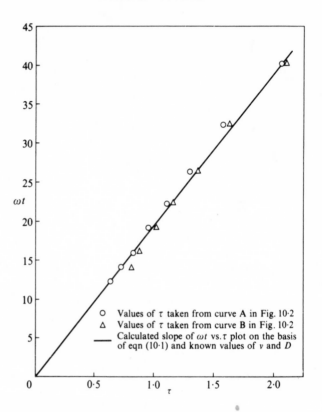

FIG. 10.3. Experimental ωt values for System I plotted against theoretical τ values. $\omega = 125{\cdot}7 \, \text{rad s}^{-1}$; positive disc current step. Electrode geometry: $r_1 = 0{\cdot}4770 \, \text{cm}$; $r_2 = 0{\cdot}4869 \, \text{cm}$; $r_3 = 0{\cdot}5222 \, \text{cm}$.

We can also use the parameters a, b, and c tabulated in Appendix 5 to calculate transients at short times. When eqn (9.10) is substituted in eqn (10.5) we obtain the general expression

$$\frac{N_\tau}{N_0} = \exp(b) \int_0^\tau f(\tau - \lambda) \left[\exp\left(-\frac{a^2}{4\lambda}\right)\left(\frac{a}{2\sqrt{(\pi \lambda^3)}} - \frac{1}{\sqrt{(\pi \lambda)}}\right) + \right.$$

$$\left. + (c+1) \exp\left\{(c+1)^2 \lambda + a(c+1)\right\} \text{erfc}\left\{(c+1)\sqrt{(\lambda)} + \frac{a}{2\sqrt{\lambda}}\right\}\right] d\lambda.$$

For a galvanostatic step at the disc $\quad f(\tau - \lambda) = -1$

and $\qquad \dfrac{N_\tau}{N_0} = \dfrac{\exp(b)}{c+1} \times$

$$\times \left[\exp\left\{(c+1)^2 \tau + a(c+1)\right\} \text{erfc}\left\{(c+1)\sqrt{(\tau)} + \frac{a}{2\sqrt{\tau}}\right\} + c\, \text{erfc}\left(\frac{a}{2\sqrt{\tau}}\right)\right]$$

$$\text{(10.9)}$$

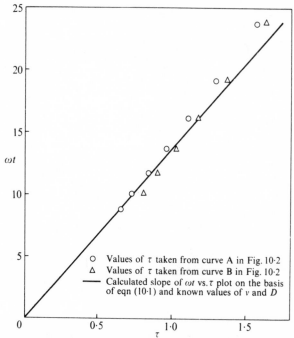

FIG. 10.4. Experimental ωt values for System II plotted against theoretical τ values. $\omega = 125 \cdot 7 \, \text{rad s}^{-1}$; negative disc current step.

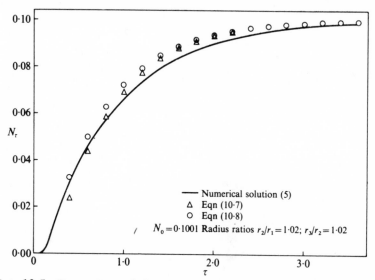

FIG. 10.5. Comparison of theoretical ring current transients calculated numerically and analytically.

TABLE 10.1

Values of ωt for System I

ω rad s^{-1} Step N_t/N_0	62·8		125·7		251·4		Eqn (10.7)	Eqn (10.8)
	+	−	+	−	+	−		
0·938	40·5	40·2	40·2	40·2	40·4	40·4	40·2	40·7
0·875	32·4	32·1	32·4	32·4	32·2	32·2	≲30·5	31·6
0·812	26·7	26·7	26·4	26·7	26·4	26·4	25·2	26·6
0·750	22·6	22·3	22·3	22·6	22·3	22·3	21·5	22·9
0·687	19·2	19·2	19·2	19·2	19·3	19·3	18·6	20·0
0·625	16·0	16·3	16·0	16·0	16·3	16·3	16·3	17·5
0·563	14·1	14·1	14·1	14·1	14·3	14·3	14·2	15·7
0·500	12·2	12·6	12·2	12·6	12·3	12·3	12·6	−

TABLE 10.2

Values of ωt for System II

ω rad s^{-1} Step N_t/N_0	62·8		125·7		251·4		Eqn (10.7)	Eqn (10.8)
	+	−	+	−	+	−		
0·875	22·6	22·4	23·6	23·8	23·6	23·7	21·2	22·1
0·812	18·7	18·8	19·1	19·2	19·4	19·2	17·6	18·6
0·750	15·7	15·7	16·0	16·2	16·3	16·0	15·0	16·0
0·687	13·8	13·4	14·1	13·7	14·3	13·9	13·0	14·0
0·625	11·9	11·3	12·1	11·8	12·2	11·9	11·4	12·2
0·563	10·7	9·9	10·6	10·2	10·7	10·2	9·9	11·0
0·500	9·1	8·6	9·3	8·8	9·5	9·0	8·8	−

A typical result from this equation for short times is shown in Fig. 10.2, and it can be seen that it meets up with eqns (10.7) and (10.8).

The Laplace Technique

As can be seen even for this simple case the inversion of the transforms back into real time is not easy, and leads to rather complex series solutions. Also the inversion can only be carried out directly for $\tau \sim$ or > 1. In order to obtain information at smaller values of τ and for more complicated systems than a simple step function we have developed a small analogue device which carries out the Laplace transformation of the ring current. This quantity $\overset{\mathsf{\scriptscriptstyle WW}}{i}_R$ where

$$\overset{\mathsf{\scriptscriptstyle WW}}{i}_{R,t} - \overset{\mathsf{\scriptscriptstyle WW}}{i}_{R,0} = \int_0^\infty (i_{R,t} - i_{R,0}) e^{-st} \, dt$$

can be interpreted using eqns (10.4) and (10.5). Thus instead of carrying out the theoretical inversion we transform the experimental data. The time scale for the ring current transient ($\sim 10^{-1}$ s) is convenient for the direct transformation. It is not so short that one has problems with high speed relays, 'ringing', etc., nor so long that one has problems of stability and the integration of noise. A block diagram of the unit is shown in Fig. 10.6. A more detailed description

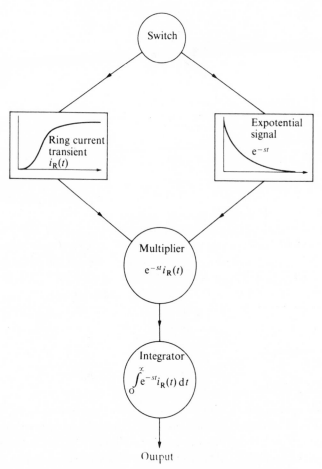

FIG. 10.6. Block diagram of the Laplace Unit.

has been published.[4] The whole unit only costs about £150 in components. The value of s in the above equation is fixed by an RC network and so one measures $\overset{\wedge\wedge\wedge}{i_R}$ as a function of s. Equations (10.4) and (10.5) give

$$\bar{N}_T = -\frac{(\bar{i}_{R,T} - \bar{i}_{R,0})}{i_{D,\infty} - i_{D,0}} = -N_{\sigma 1/2}\,\bar{f}(\sigma),$$

where $^-$ represents transformation with respect to T and $\wedge\wedge\wedge$ with respect to t.

Conversion between the two variables is made[4] by

$$\overset{\wedge\wedge\wedge}{F_t} = \frac{t}{T}\cdot\bar{F}_T \quad \text{and} \quad \sigma T = st, \tag{10.10}$$

where

$$t/T = C^{-2/3}D^{-1/3}.$$

Then

$$\overset{\wedge\wedge\wedge}{N_t} = -\frac{\overset{\wedge\wedge\wedge}{i_{R,t}} - \overset{\wedge\wedge\wedge}{i_{R,0}}}{i_{D,\infty} - i_{D,0}} = -C^{-2/3}D^{-1/3}N_{\sigma 1/2}\,\bar{f}(\sigma),$$

or

$$\bar{f}(\sigma) = \frac{C^{2/3}D^{1/3}}{N_{\sigma 1/2}} \times \frac{\overset{\wedge\wedge\wedge}{i_{R,t}} - \overset{\wedge\wedge\wedge}{i_{R,0}}}{i_{D,\infty} - i_{D,0}}, \tag{10.11}$$

where $N_{\sigma 1/2}$ is N_K with κ replaced by $\sigma^{1/2}$.

For the particular case of a step in the disc current $\bar{f}(\sigma) = -1/\sigma$ and

$$\overset{\wedge\wedge\wedge}{N_t} = C^{-2/3}D^{-1/3}N_{\sigma 1/2}\,\sigma^{-1} = N_{\sigma 1/2}\,s^{-1}. \tag{10.12}$$

The left-hand side of this equation can be measured directly using the Laplace technique. The right-hand side is calculated, by using eqn (10.10) to find σ and then using the calculations in the previous chapter to find $N_{\sigma 1/2}$. We define this quantity as $\overset{\wedge\wedge\wedge}{N}_G$. A comparison between the values is made in Tables 10.3 and 10.4 for the $Fe(CN)_6^{3-}/Fe(CN)_6^{4-}$ system (System I of the previous section). Table 10.4 shows results for large values of σ corresponding to analysis of the early part of the transient. A series of e^{-st} curves plotted on the same scale as a transient are shown in Fig. 10.7. It can be seen that s as large as 100 concentrates mainly on the take off point.

The satisfactory agreement between theory and experiment shows the power of the Laplace technique. It should be pointed out that each reading of $\overset{\wedge\wedge\wedge}{i_R}$ is a steady signal read on a digital voltmeter; we do not have to try and measure current versus time data from an oscilloscope trace. The technique could well be used for most transients involving concentration polarization since the time scale of this process is convenient for direct analogue conversion.

TABLE 10.3

Experimental values, $\tilde{\tilde{N}}_t$, and theoretical values, $\tilde{\tilde{N}}_G$, of the transformed time-dependent collection efficiency

	Experimental						Theoretical	
W(Hz)	10		20		40		Eqns (9.9),	Eqns (9.9),
Step	+	–	+	–	+	–	(9.8), (9.5)	(9.8), (9.6)
W/s								
40·0	–	–	–	–	2·45	2·48	2·553	2·529
20·0	–	–	1·17	1·16	1·16	1·17	1·186	1·164
16·0	–	–	0·91	0·89	0·89	0·90	0·915	0·896
12·0	–	–	0·64	0·63	0·64	0·63	0·648	0·632
10·0	0·50	0·49	0·51	0·50	–	–	0·517	0·504
8·0	0·379	0·371	0·384	0·379	0·385	0·381	0·388	0·380
6·0	0·259	0·253	0·264	0·257	0·252	0·261	0·264	0·265
4·0	0·144	0·140	0·148	0·143	0·145	0·145	0·146	–
2·0	0·046	0·043	0·047	0·046	–	–	0·047	–
1·0	0·011	0·011	–	–	–	–	0·012	–

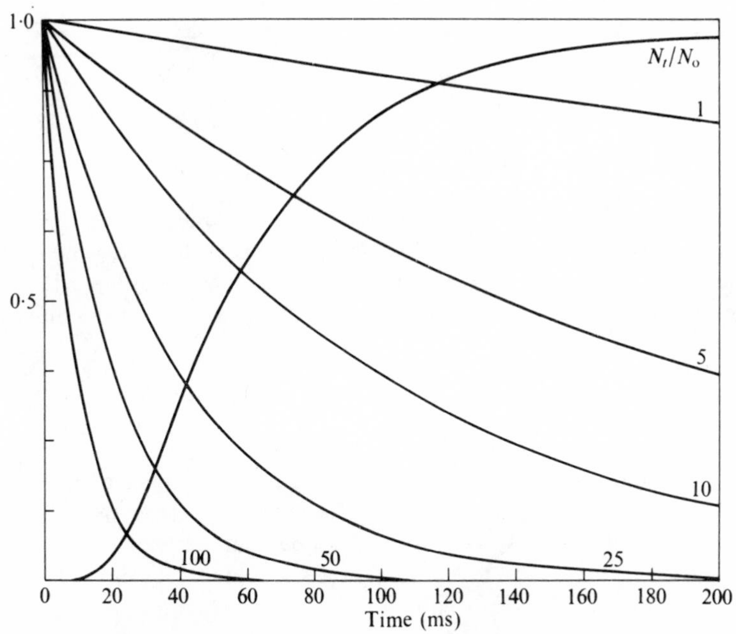

FIG. 10.7. Ring current transient and exponential decay curves. The appropriate value of s is marked on each exponential curve.

TABLE 10.4

Experimental values, N_t, and theoretical values, N_G, of the transforme time-dependent collection efficiency at short times*

	$W = 20\,\text{Hz}$				Positive step			
s	10·0	12·5	16·7	20·0	25·0	33·3	50·0	100·0
W/s	2·00	1·60	1·20	1·00	0·80	0·60	0·40	0·20
Experimental	0·0469	0·0295	0·0165	0·0108	0·0066	0·0034	0·0011	0·0002
Theoretical	0·0467	0·0308	0·0174	0·0118	0·0071	0·0035	0·0011	0·0001

Potential Step at the Disc

To describe the detailed conditions on the disc electrode it is probably better to control the potential of the electrode rather than the current. Adsorption equilibrium constants and electrochemical rate constants are then more likely to have steady values. If we step the potential, the function $\bar{f}$ does not have the simple form of $-1/\sigma$, but has the more complicated form given below,[6] which describes the disc current transient:

$$\bar{f}(\sigma) = -\frac{1 + \lambda_1 A_1}{\sigma^{1/2}\{\sigma^{1/2} + \lambda_1 \tanh(A_1 \sigma^{1/2})\}}, \qquad (10.13)$$

where

$$\lambda_1 = \frac{k_1' + k_{-1}'}{D^{2/3} C^{1/3}}$$

and k_1' and k_{-1}' are the heterogeneous rate constants for the forward and backward reactions at the particular potential to which the electrod is stepped. Three approximations for $\bar{f}$ may be obtained depending on the size of λ_1 and σ. First, for systems of low λ_1, where the electrode kinetics are slow,

$$\bar{f}(\sigma) \simeq -\frac{1}{\sigma}.$$

This is the same result as for the galvanostatic step, since for small λ_1 the concentration polarization is not important, the current is a unique function of the potential, and it does not matter which is controlled.

On the other hand, for large λ_1 we must distinguish between large and small values of σ. For large σ (corresponding to small τ)

$$\bar{f}(\sigma) \simeq -\lambda_1 A_1 \sigma^{-1},$$

and
$$f(\tau) \simeq -\lambda_1 A_1, \quad \lambda_1 \gg 1. \tag{10.14}$$

For $\lambda_1 > \sigma^{1/2} > 1$:
$$\bar{f}(\sigma) \simeq -A_1/\sigma^{1/2},$$

and
$$f(\tau) \simeq -A_1(\pi\tau)^{-1/2}, \quad \lambda_1 \gg \tau^{-1/2} \gg 1. \tag{10.15}$$

And for $1 > \sigma^{1/2}$:
$$\bar{f}(\sigma) \simeq -1/\sigma,$$
$$f(\tau) \simeq -1. \tag{10.16}$$

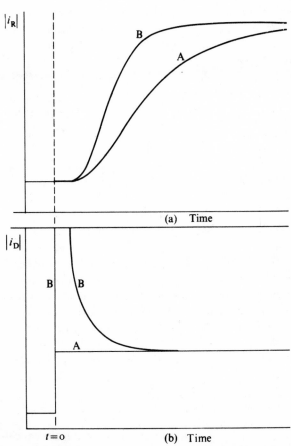

FIG. 10.8. Disc and ring currents for current and potential steps at the disc. (a) Ring current against time. A, Ring current transient for current step at the disc. B, Ring current transient for potential step at the disc. (b) Disc current against time. A, Current step. B, Disc current for potential step.

F

Thus $f(\tau)$ starts large and decreases. Equation (10.14) describes the initial control by the electrode reaction; eqn (10.15) describes the concentration polarization beginning to control the reaction and the spread of the concentration gradients out into the stagnant diffusion layer; finally, eqn (10.16) describes the steady state.

If eqn (10.13) for $\bar{f}$ is put into eqn (10.11) we obtain

$$\overset{\text{\tiny WW}}{N}_t = \frac{C^{-2/3}D^{-1/3}\,N_{\sigma^{1/2}}\,(1 + \lambda_1 A_1)}{\sigma^{1/2}\{\sigma^{1/2} + \lambda_1\,\tanh(A_1\,\sigma^{1/2})\}}. \qquad (10.17)$$

It would not, to say the least, be easy to invert this function into real time. However, using the Laplace technique we do not have to perform this step. Fig. 10.8 shows schematically the difference in the disc and ring current transients for a current and a potential step. The potential step produces a larger current at $t = 0$ which decays as concentration polarization is established. This leads to a ring current transient that rises more steeply.

Equation (10.17), which describes the Laplace transform of the ring current transient, was tested using the Laplace technique for the $Fe(CN)_6^{3-}/Fe(CN)_6^{4-}$ system. The equation is rearranged:

$$\frac{\overset{\text{\tiny WW}}{N}_G}{\overset{\text{\tiny WW}}{N}_t} = \frac{1}{1 + \lambda_1 A_1} + \frac{\lambda_1}{1 + \lambda_1 A_1}\,\frac{\tanh(A_1\,\sigma^{1/2})}{\sigma^{1/2}}, \qquad (10.18)$$

where from eqn (10.12) $N_G = s^{-1}N_{\sigma^{1/2}}$ and is the equivalent galvanostatic transient.

Fig. 10.9 shows a plot of $\overset{\text{\tiny WW}}{N}_G/\overset{\text{\tiny WW}}{N}_t$ against $\tanh(A_1\sigma^{1/2})/\sigma^{1/2}$.[7] Reasonable straight lines are obtained for each rotation speed. Equation (10.18) predicts that for $\sigma = 0$ all lines pass through the point $(1\cdot0, 1\cdot288)$ since $A_1 = 1\cdot288$. The gradients and intercepts depend on the size of λ_1. For $\lambda_1 \ll 1$ we have an irreversible system with very little concentration polarization and, as discussed above, we obtain the same transient behaviour at the disc whether the current or the potential is controlled. So $N_t = N_G$ as shown by the dotted line in Fig. 10.9. On the other hand, for $\lambda_1 \gg 1$ we have a reversible system, and eqn (10.18) reduces to the other limiting case

$$\frac{\overset{\text{\tiny WW}}{N}_G}{\overset{\text{\tiny WW}}{N}_t} = \frac{\tanh(A_1\,\sigma^{1/2})}{A_1\,\sigma^{1/2}}$$

Values of $\lambda_1 \sim 1$ should give straight lines which lie between these two extremes.

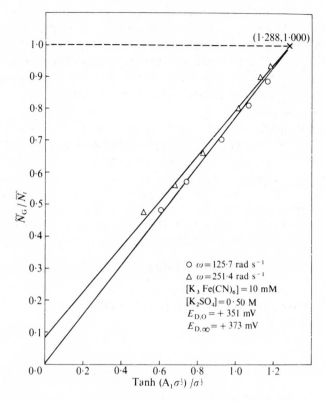

FIG. 10.9. Test of eqn (10.18).

Adsorption, Double-Layer Charging, and Other Effects

So far in the discussion of transients we have assumed that the
disc current and the flux of intermediate are directly proportional at
all times. However, although this must be true for steady-state
observations it is by no means necessarily the case for transient
studies. Reasons for deviations from direct proportionality are, for
instance, the charging of the double layer, adsorption of the electrogenerated intermediate, change in the state of oxidation of the electrode, and changes in the surface concentration of intermediates in the
electrode reaction.

From eqn (10.11),

$$\bar{f}(\sigma) = \frac{C^{2/3} D^{1/3}}{N_{\sigma^{1/2}}} \times \frac{\widetilde{\widetilde{i}}_{R,t} - \widetilde{\widetilde{i}}_{R,0}}{i_{D,\infty} - i_{D,0}} \, .$$

Since all the quantities on the right-hand side can be measured, a
calculated $\bar{f}(\sigma)$ can be measured from the ring-current transient.

Thus a study of ring-disc electrode transients gives one independent measurements of the flux of electrons, the disc current i_D, and the flux of material at the disc surface, the function $\bar{f}(\sigma)$. Hence at any time in the transient one can measure the proportion of the current that is actually producing intermediate in the solution. This technique is likely to be very powerful for the elucidation of the details of electrode processes.

For the use of the current step at the disc we can list some different forms of $\bar{f}$ that depend on the processes taking place on the disc electrode.

(1) No complications: $\bar{f} = -1/\sigma$.

(2) Double layer charging:
We assume an equivalent circuit of a capacitor C_{DL} and charge transfer resistance R_{CT} in parallel, then

$$\bar{f} = -1/\sigma(1 + C_{DL} R_{CT} C^{2/3} D^{1/3} \sigma).$$

(3) Adsorption of the intermediate:
If we assume that the intermediate is adsorbed on the disc according to

$$c_{ads} = K c_{w=0}$$

then

$$\bar{f} = -1/\sigma\{1 + KC^{1/3} D^{-1/3} \sigma^{1/2} \tanh(A_1 \sigma^{1/2})\}. \quad (10.19)$$

(4) Two-step electrode mechanism:
We assume an electrode mechanism of the type

$$
\begin{array}{ccc}
\text{solution} & \text{A} & \text{C} \\
 & \Updownarrow & \Updownarrow \\
\text{electrode} & \text{A}_{Ads} \xrightarrow{k_1''} \text{B}_{Ads} \xrightarrow{k_2''} \text{C}_{Ads} \\
 & a' \quad\quad b' \quad\quad c'
\end{array}
$$

in which a' and k_1''/k_2'' remain constant; k_1'' and k_2'' represent two n-electron transfers with the same transfer coefficient.
Then

$$b' = (k_1''/k_2'') a' (1 - u)$$

where

$$u = \exp\{-k_2'' t - \tfrac{1}{2}(1 - u)\}$$

$$\simeq \exp(-k_2'' t - \tfrac{1}{2}) \quad \text{for} \quad k_2'' t > 1$$

and

$$\bar{f} = -\frac{1 - \sigma \exp(-\tfrac{1}{2})/(\sigma + k_2'' C^{-2/3} D^{-1/3})}{\sigma\{1 + KC^{1/3} D^{-1/3}\sigma^{1/2} \tanh(A_1 \sigma^{1/2})\}} . \quad (10.20)$$

Similar expressions can be obtained for cases where the potential of the disc electrode is controlled. Under these conditions the disc current transient must also be measured, and a comparison between theory and experiment can be made for both the disc current and the flux of material produced. The differences between the two may be caused, for instance, by double-layer charging, adsorption, the generation or destruction of an electrode intermediate, and the oxidation or reduction of the electrode surface.

An estimate of the sensitivity of the technique for measuring adsorption and the lifetime of intermediates may be made from eqns (10.19) and (10.20).

For adsorption, eqn (10.19), it is necessary for $K C^{1/3} D^{-1/3} > \sim 1$, but $C^{1/3} D^{-1/3} \sim 10^3 \text{cm}^{-1}$ and therefore $K > \sim 10^{-3} \text{cm}$. If we can measure currents from concentrations as low as 10^{-6}M or $10^{-9} \text{mol cm}^{-3}$ then one should be able to measure changes in coverage as low as $10^{-11} \text{mol cm}^{-2}$. For the lifetime of an intermediate [eqn (10.20)]

$$k_2'' C^{2/3} D^{-1/3} < \sim 3,$$

where $C^{-2/3} D^{-1/3} \sim 0.1 \text{s}$ and thus $k_2'' \sim 30 \text{s}^{-1}$. Hence the lifetime of moderately unstable intermediates in the electrode mechanism should be able to be measured, even if they are not desorbed from the electrode. The measurement of $\bar{f}$ and its interpretation, using equations as above, should allow the quantitative interpretation of the various parameters. Already the ring-disc electrode has been applied successfully by Bruckenstein and Napp[8] to a study of the adsorption of Cu(II) on Pt in 0.5 M HCl.

Integration of Transients

Another important application of the theory of transients is in the measurement of the total quantity of material deposited on or released by a disc electrode. For instance, imagine that the disc electrode is an alloy undergoing electrochemical dissolution; one component, A, of the alloy can be detected at the ring electrode. The ring current is integrated with time and since at $t = 0$, $i_D = i_R = 0$ and

$$i_{D,t\to\infty} = \pi r_1^2 n_D F C^{1/3} D^{2/3} (\partial c/\partial w)_{\substack{w=0 \\ r<r_1 \\ t\to\infty}},$$

from eqns (10.3) and (10.5)

$$\int_0^t i_R \, dt = -\pi r_1^2 n_R F C^{1/3} D^{2/3} \mathcal{L}^{-1} \{\sigma^{-1} N_K \mathcal{L}(\partial c/\partial w)_{\substack{w=0 \\ r<r_1 \\ t}}\},$$

where κ in N_κ is replaced by $\sigma^{1/2}$. As $t \to \infty$, $\sigma \to 0$ and $N_\kappa \to N_0$. Also, if c_T is the total number of moles of A dissolved from the disc,

$$c_T = -\pi r_1^2 C^{1/3} D^{2/3} \int_0^\infty (\partial c/\partial w)_{\substack{w=0 \\ r<r_1 \\ t}} dt,$$

and we obtain the simple result

$$c_T = (N_0 n_R F)^{-1} \int_0^\infty i_R dt.$$

This result holds whatever the transient behaviour of i_R. It is not necessary for $i_{R,t=0} = 0$. As long as $i_R = i_{R,c} + i_{R,0}$ where $i_{R,0}$ is constant with time and $i_{R,c}$ is caused by c then:

$$c_T = (N_0 n_R F)^{-1} \int_0^\infty (i_R - i_{R,0}) dt.$$

Furthermore, this equation will hold for the deposition of A on the electrode when $i_R < i_{R,0}$ and c_T is negative. Since with $i_R \sim 10^{-5} A$,

$$t \sim 10^{-2} s$$

and

$$F \sim 10^5 C s^{-1} mol^{-1},$$

$$c_T \sim 10^{-12} mol,$$

very small amounts of adsorption, deposition or dissolution can be detected if the processes at the disc electrode are rapid enough.

This conclusion has been tested using square and triangular pulses for the $Fe(CN)_6^{3-}/Fe(CN)_6^{4-}$ system. Results are given in Table 10.5.

TABLE 10.5

Integration of transients

Pulse	$\int i_D dt$ mC	$\int i_R dt$ mC	$\int i_R dt / \int i_D dt$	Steady-state collection efficiency N_0
Saw tooth	0.43_0	93.6	0.218	0.219
	0.52_5	114.0	0.217	0.219
	0.67_4	147.0	0.218	0.219
Square	1.18_0	238.0	0.202	0.204
	0.92_7	185.0	0.200	0.204
	0.81_3	163.0	0.200	0.204

Alternating Current Transients

Very recently we have been able to develop the theory of a.c. transients at the ring-disc electrode.[6] If an alternating current is passed through the disc electrode then an alternating current is seen on the ring. The amplitude and phase of the ring current will be shifted with respect to the disc current and will be a function of the frequency. For very low-frequency currents the phase shift will be zero and the amplitudes will be related by the steady-state collection efficiency N_0. On the other hand, for very high-frequencies the a.c. signal will not be seen on the ring electrode and the amplitude will be zero.

The theory describing this starts from eqn (10.5):

$$N_T = -N_\sigma 1/2 \, \bar{f}(\sigma).$$

We now write

$$f(\tau) = -\cos \omega' \tau$$
$$= -\cosh "i" \omega' \tau,$$

where $\omega' = C^{-2/3} D^{-1/3} 2\pi fr$ and fr is the frequnecy in Hz; $"i" = \sqrt{-1}$ and is written thus to avoid confusion with i, a current.

By the convolution theorem

$$N_T = \frac{1}{2} \int_0^T [\exp\{"i"\omega'(\tau - \chi)\} + \exp\{-"i"\omega'(\tau - \chi)\}] F(\chi) d\chi,$$

where $F(\chi) = \mathcal{L}^{-1} N_K(\chi)$; that is N_K with κ replaced by $\chi^{1/2}$.

Let the a.c. signal be applied for long enough so that initial transient effects are removed. Then we can let $\tau \to \infty$ and

$$
\begin{aligned}
N_{a.c.} &= \tfrac{1}{2} \exp("i"\omega'\tau) \int_0^\infty \exp(-"i"\omega'\chi) F(\chi) d\chi + \\
&\quad + \tfrac{1}{2} \exp(-"i"\omega'\tau) \int_0^\infty \exp("i"\omega'\chi) F(\chi) d\chi \\
&= \tfrac{1}{2} \exp("i"\omega'\tau) N_K("i"\omega') + \tfrac{1}{2}\exp(-"i"\omega'\tau) N_K(-"i"\omega') \\
&= Y \cos \omega' \tau - X \sin \omega' \tau
\end{aligned}
$$

where

$$N_K("i"\omega') = Y + X "i". \tag{10.21}$$

Thus the problem of describing the phase shift, δ, and amplitude, $N_{a.c.}$ of the ring current with respect to the disc current reduces to replacing χ in the expression for N_K with $\sqrt{"i"\omega'}$ and then working out the real and imaginary parts of N_K. We have done this using the computer programme which inverts eqn (9.7) and using the correction in eqn (9.9). The results are compared with experiments[9] for the $Fe(CN)_6^{4-}/Fe(CN)_6^{3-}$ system in Fig. 10.10. Good agreement is found.

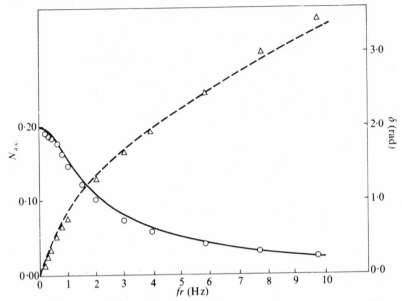

FIG. 10.10. Amplitude and Phase Shift for a.c. transients.

For high frequencies ($\omega' > 1$) one may also use the parameters a, b, and c given in Appendix 5. The quantities Y and X in eqn (10.21) can then be calculated directly:

$$Y = X'\left[Y'\cos\left(a\sqrt{\frac{\omega'}{2}}\right) + \sqrt{\frac{\omega'}{2}}\sin\left(a\sqrt{\frac{\omega'}{2}}\right)\right]$$

$$X = X'\left[\sqrt{\frac{\omega'}{2}}\cos\left(a\sqrt{\frac{\omega'}{2}}\right) - Y'\sin\left(a\sqrt{\frac{\omega'}{2}}\right)\right],$$

where

$$X' = \frac{N_0\,\exp(b)\,\exp\left(-a\sqrt{\frac{\omega'}{2}}\right)}{\left(1 + c + \sqrt{\frac{\omega'}{2}}\right)^2 + \frac{\omega'}{2}}$$

and

$$Y' = \left(c + \sqrt{\frac{\omega'}{2}}\right)\left(1 + c + \sqrt{\frac{\omega'}{2}}\right) + \frac{\omega'}{2}.$$

For lower values of ω' we can use the expression given in (9.8) for N_K.

The successful description of alternating currents at ring-disc electrodes is particularly useful since from the a.c. signal observed on the ring electrode we can now obtain directly the phase and amplitude shift in the production of material at the disc electrode as a function of the a.c. frequency. This allows us to construct directly the R.C. analogue for the electrode reaction.

In conclusion, let us emphasize once again that the study of transients on a ring-disc electrode is a unique technique for the investigation of the mechanism of electrode reactions, since one measures separately the flux of electrons (i_D) and the flux of material (i_R). To return to our first quotation, although the currents are distinctly fluctuating at least they are reasonably reproducible.

References

1. ALBERY, W.J. (1967) *Trans. Faraday Soc.* 63, 1771.

2. MCKAY, A.T. (1930) *Proc. phys. Soc.* 42, 547.

3. HALE, J.M. (1965) *In Batteries*, Vol. 2, p. 147. Pergamon Press, Oxford.

4. ALBERY, W.J., DRURY, J.S., and HITCHMAN, M.L. (1970) *Trans. Faraday Soc.* In press.

5. BARD, A.J. and PRATER, K.B. (1970) *J. electrochem. Soc.* 117, 207.

6. ALBERY, W.J. (1970) Unpublished results.

7. ALBERY, W.J. and DRURY, J.S. (1970) Unpublished results.

8. BRUCKENSTEIN, S. and NAPP, D.T. (1968) *J. Am. chem. Soc.* 90, 6303.

9. ALBERY, W.J. and DRURY, J.S. (1970) Unpublished results.

APPENDIX 1

THIS appendix contains values of the functions $F(\theta)$ and $G(\phi)$ described in Chapter 3, and values of the collection efficiency N_0 for common radius ratios.

TABLE 1

Values of $F(\theta) - eqn$ (3.13)

θ	·00	·01	·02	·03	·04	·05	·06	·07	·08	·09
0	·0000	·1777	·2234	·2551	·2801	·3010	·3191	·3351	·3495	·3627
·10	·3748	·3860	·3964	·4063	·4155	·4242	·4325	·4403	·4478	·4550
·20	·4619	·4685	·4748	·4809	·4867	·4924	·4978	·5031	·5082	·5132
·30	·5180	·5227	·5272	·5316	·5359	·5400	·5441	·5480	·5519	·5557
·40	·5593	·5629	·5664	·5698	·5732	·5765	·5797	·5828	·5859	·5889
·50	·5918	·5947	·5975	·6003	·6030	·6057	·6083	·6109	·6135	·6159
·60	·6184	·6208	·6231	·6255	·6278	·6300	·6322	·6344	·6365	·6386
·70	·6407	·6427	·6447	·6467	·6487	·6506	·6525	·6544	·6562	·6580
·80	·6598	·6616	·6633	·6650	·6667	·6684	·6700	·6717	·6733	·6749
·90	·6764	·6780	·6795	·6810	·6825	·6840	·6854	·6869	·6883	·6897

TABLE 2

Values of $G(\phi)$ *– eqn* (3.14)

ϕ	·00	·01	·02	·03	·04	·05	·06	·07	·08	·09
0	·0000	·0191	·0302	·0395	·0476	·0550	·0619	·0683	·0744	·0802
·10	·0857	·0910	·0961	·1010	·1057	·1103	·1148	·1191	·1233	·1274
·20	·1313	·1352	·1390	·1427	·1463	·1499	·1533	·1567	·1600	·1633
·30	·1665	·1696	·1727	·1757	·1787	·1816	·1845	·1873	·1901	·1929
·40	·1955	·1982	·2008	·2034	·2059	·2084	·2109	·2133	·2157	·2181
·50	·2204	·2227	·2250	·2272	·2294	·2316	·2338	·2359	·2380	·2401
·60	·2421	·2442	·2462	·2482	·2501	·2521	·2540	·2559	·2578	·2596
·70	·2614	·2633	·2651	·2668	·2686	·2703	·2721	·2738	·2755	·2771
·80	·2788	·2804	·2821	·2837	·2853	·2868	·2884	·2900	·2915	·2930
·90	·2945	·2960	·2975	·2990	·3004	·3019	·3033	·3047	·3061	·3075

TABLE 3

Values of N_0 *for common radius ratios*

r_3/r_2	r_2/r_1								
	1·02	1·03	1·04	1·05	1·06	1·07	1·08	1·09	1·10
1·02	·1013	·0976	·0947	·0922	·0902	·0884	·0869	·0855	·0843
1·03	·1293	·1250	·1215	·1186	·1162	·1140	·1121	·1104	·1089
1·04	·1529	·1483	·1444	·1412	·1385	·1360	·1339	·1320	·1302
1·05	·1737	·1687	·1647	·1612	·1582	·1556	·1533	·1512	·1493
1·06	·1923	·1872	·1829	·1793	·1761	·1733	·1708	·1686	·1665
1·07	·2092	·2039	·1996	·1958	·1925	·1896	·1869	·1846	·1824
1·08	·2247	·2194	·2149	·2110	·2076	·2046	·2019	·1994	·1972
1·09	·2392	·2338	·2292	·2252	·2217	·2186	·2158	·2133	·2110
1·10	·2526	·2472	·2426	·2385	·2350	·2318	·2289	·2263	·2240
1·12	·2772	·2717	·2670	·2629	·2593	·2560	·2530	·2503	·2479
1·14	·2992	·2938	·2890	·2849	·2812	·2778	·2748	·2720	·2695
1·16	·3192	·3138	·3090	·3048	·3011	·2977	·2947	·2919	·2893
1·18	·3375	·3321	·3274	·3232	·3194	·3161	·3130	·3101	·3075
1·20	·3544	·3490	·3443	·3402	·3364	·3330	·3299	·3271	·3245
1·22	·3701	·3648	·3601	·3560	·3523	·3489	·3458	·3429	·3403
1·24	·3848	·3795	·3749	·3708	·3671	·3637	·3606	·3577	·3551
1·26	·3985	·3933	·3887	·3847	·3810	·3776	·3745	·3717	·3691
1·28	·4115	·4063	·4018	·3977	·3941	·3907	·3877	·3849	·3822
1·30	·4237	·4186	·4141	·4101	·4065	·4032	·4001	·3973	·3947
1·32	·4353	·4302	·4258	·4218	·4183	·4150	·4119	·4092	·4066
1·34	·4463	·4413	·4369	·4330	·4294	·4262	·4232	·4204	·4178
1·36	·4567	·4518	·4475	·4436	·4401	·4369	·4339	·4311	·4286
1·38	·4667	·4619	·4576	·4538	·4503	·4471	·4441	·4414	·4389
1·40	·4762	·4715	·4673	·4635	·4600	·4568	·4539	·4512	·4487

APPENDIX 2

THIS appendix considers the case of linked chemical equilibria next to the electrode.

First we take the case of $n - 1$ equilibria of which the jth equilibrium is sluggish. We assume that for all the others we can write (using the notation of Chapter 5)

$$c_{p+1} = K_p c_p,$$

where we have dropped the second subscript, q.

Considering each species

$$k_{p-1} \Big\Updownarrow k_{-(p-1)}$$

$$A_p$$

$$k_p \Big\Updownarrow k_{-p}$$

we can write down the diffusion/kinetic equation

$$\frac{D \partial^2 c_p}{\partial z^2} = k_{p-1} c_{p-1} - k_{-(p-1)} c_p - k_p c_p + k_{-p} c_{p+1}.$$

Adding together all the equations from 1 to j and all the equations from $j + 1$ to m, and then using the $(m - 2)$ 'equilibrium' relationships we obtain

$$D \left(1 + \frac{1}{K_{j-1}} + \frac{1}{K_{j-1} K_{j-2}} + \ldots\ldots + \frac{1}{K_{j-1} \ldots\ldots K_1} \right) \frac{\partial^2 c_j}{\partial z^2}$$

$$= k_j c_j - k_{-j} c_{j+1}, \tag{A2.1}$$

and

$$D \{ 1 + K_{j+1} + K_{j+1} K_{j+2} + \ldots\ldots + (K_{j+1} \ldots\ldots K_{m-1}) \} \frac{\partial^2 c_{j+1}}{\partial z^2}$$

$$= -k_j c_j + k_{-j} c_{j+1}. \tag{A2.2}$$

These equations describe the concentration profiles in the region of the thickest reaction layer corresponding to the least labile equilibrium. Between the inside of this reaction layer and the electrode there will be thinner reaction layers for the more labile equilibria. If eqns (A2.1) and (A2.2) hold for $z > z^*$, where z^* describes the boundary between the thin reaction layers and the thick one, then it is not necessarily true that

$$g_{j, z=0} = g_{j, z=z^*}.$$

In fact,

$$\sum_{l=1}^{l=j} g_{l, z=0} = \sum_{l=1}^{l=j} g_{l, z=z^*}.$$

This equation describes a constant flux with distance of all the species C_1 to C_j in the region $z = 0$ to $z = z^*$. Since z^* is smaller than the reaction layer for $C_j \rightleftharpoons C_{j+1}$ there is little transformation between C_j and C_{j+1} in this region. Since 'equilibrium' through C_1 to C_j is established at $z = z^*$,

$$\sum_{l=1}^{l=j} g_{l, z=z^*} = g_{j, z=z^*} S_1,$$

where S_1 is the series on the left of eqn (A2.1).
Therefore

$$g_{j, z=z^*} = \sum_{l=1}^{l=j} g_l / S_1.$$

Similarly

$$g_{j+1, z=z^*} = \sum_{l=j+1}^{l=m} g_l / S_2,$$

where S_2 is the series on the left of eqn (A2.2).

Following a similar argument to that used for eqn (5.7) we obtain from eqns (A2.1) and (A2.2) the relationship

$$\frac{K_j \sum_{l=1}^{l=j} g_l}{S_1} - \frac{\sum_{l=j+1}^{l=m} g_l}{S_2} = -\frac{Z_D}{\mu} (K_j c_j - c_{j+1}), \tag{A2.3}$$

where

$$\mu = \sqrt{\frac{D}{\dfrac{k_j}{S_1} + \dfrac{k_{-j}}{S_2}}}$$

We now give the full treatment, including allowance for convective-diffusion, for the system of two linked chemical equilibria

$$A \underset{k_{-1}}{\overset{k_1}{\rightleftharpoons}} B \underset{k_{-2}}{\overset{k_2}{\rightleftharpoons}} C.$$

We write the basic equations as

$$D \frac{\partial^2 a}{\partial z^2} - k_1 a + k_{-1} b = 0, \tag{A2.4}$$

$$D \frac{\partial^2 b}{\partial z^2} + k_1 a - k_{-1} b + k_{-2} c - k_2 b = 0, \tag{A2.5}$$

$$D \frac{\partial^2 c}{\partial z^2} + k_2 b - k_{-2} c = 0. \tag{A2.6}$$

We are going to allow for the convection by applying the boundary condition that at $z = Z_D$, a, b, and c equal their bulk values. By adding the three equations and integrating we obtain the simple transport equation

$$\left\{ \frac{\partial (a + b + c)}{\partial z} \right\}_0 = \frac{(a + b + c)_\infty - (a + b + c)_0}{Z_D}.$$

The two chemical equations will be obtained in terms of the variables θ and ϕ which describe the perturbation to the equilibrium condition of each chemical step:

$$\theta = K_1 a - b \quad \text{and} \quad \phi = K_{-2} c - b,$$

where

$$K_1 = k_1/k_{-1} \quad \text{and} \quad K_{-2} = k_{-2}/k_2.$$

The distance variable is normalized with Z_D so that $x = z/Z_D$, and we define

$$\alpha = (k_1 + k_{-1}) Z_D^2/D,$$

and

$$\beta = (k_2 + k_{-2}) Z_D^2/D.$$

Multiplying eqn (A2.4) by K_1 and subtracting eqn (A2.5) gives:-

$$\frac{\partial^2 \theta}{\partial x^2} - \alpha \theta - \frac{\beta}{1 + K_{-2}} \phi = 0. \tag{A2.7}$$

Similarly, multiplication of eqn (A2.6) by K_{-2} and subtraction of eqn (A2.5) gives

$$\frac{\partial^2 \phi}{\partial x^2} - \beta \phi - \frac{\alpha}{1 + K_1} \theta = 0. \tag{A2.8}$$

The boundary conditions at $x = 0$ may be worked out from the surface concentrations and gradients to give θ_0, θ_0', ϕ_0, and ϕ_0'. At $x = 1$ ($z = Z_D$) the equilibria are all in balance in the bulk of the solution and:-

$$\theta = \phi = 0.$$

We now define two parameters λ and χ which describe two reaction layer thicknesses. They are related to α and β but are not necessarily the same because of the linking of the chemical steps:

$$\lambda^2 + \chi^2 = \alpha + \beta$$

and

$$\lambda^2 \chi^2 = \left\{ 1 - \frac{1}{(1 + K_1)(1 + K_{-2})} \right\}.$$

If α and β are very unequal then (taking $\beta > \alpha$) approximate solutions are

$$\lambda^2 = \beta + \frac{\alpha}{(1 + K_1)(1 + K_{-2})} \qquad (A2.9)$$

and

$$\chi^2 = \alpha \left\{ 1 - \frac{1}{(1 + K_1)(1 + K_{-2})} \right\}, \qquad (A2.10)$$

but when α and β are comparable then λ^2 and χ^2 are roots of the quadratic equation and the kinetic properties of one equilibrium affects the other.

Multiplying eqn (A2.8) by Y and adding it to eqn (A2.7) we obtain

$$\frac{\partial^2 \theta}{\partial x^2} + Y \frac{\partial^2 \phi}{\partial x^2} = \chi^2 (\theta + Y \phi),$$

where

$$Y = \frac{(1 + K_1)(\beta - \lambda^2)}{\alpha} = \frac{(1 + K_1)(\chi^2 - \alpha)}{\alpha}$$

from which

$$\theta + Y \phi = (\theta_0 + Y \phi_0) \cosh(\chi x) + \frac{(\theta_0' + Y \phi_0') \sinh(\chi x)}{\chi}.$$

At $x = 1$, $\theta = \phi = 0$, so

$$\frac{\tanh \chi}{\chi} = -\frac{\theta_0 + Y \phi_0}{\theta_0' + Y \phi_0'}. \qquad (A2.11)$$

Similarly we obtain

$$\frac{\tanh \lambda}{\lambda} = -\frac{\phi_0 + X \theta_0}{\phi_0' + X \theta_0'}, \qquad (A2.12)$$

where

$$X = \frac{(1 + K_{-2})(\beta - \lambda^2)}{\beta} = \frac{(1 + K_{-2})(\chi^2 - \alpha)}{\beta}.$$

The allocation of the roots λ and χ is unimportant since it may be shown that

$$\frac{(1 + K_1)(\lambda^2 - a)}{a} \times \frac{(1 + K_{-2})(\chi^2 - a)}{\beta} = 1,$$

and thus changing χ into λ in eqn (A2.11) gives

$$\frac{\tanh \lambda}{\lambda} = -\frac{\theta_0 + \phi_0/X}{\theta_0' + \phi_0'/X} = -\frac{\phi_0 + X\theta_0}{\phi_0' + X\theta_0'},$$

which is the same as eqn (A2.12).

Hence these two equations are the equations describing the linked chemical equlibria , together with convective-diffusion.

We can generate eqn (A2.3) from these equations. The assumption that the thickness of each reaction layer is smaller than the diffusion layer means that a and β, and hence λ and χ, will both be much larger than 1. Hence

$$\frac{\tanh \lambda}{\lambda} = \frac{1}{\lambda} \quad \text{and} \quad \frac{\tanh \chi}{\chi} = \frac{1}{\chi}.$$

If we assume that one equilibrium (take K_2) is labile, and one (K_1) is sluggish, then a and β are well separated and we may use the approximations given in eqns (A2.9) and (A2.10). Substitution in eqns (A2.7) and (A2.6) gives

$$\theta_0 + \frac{\theta_0'}{\chi} - \left(\phi_0 + \frac{\phi_0'}{\chi}\right)\frac{1}{1 + K_{-2}} = 0 \qquad (A2.13)$$

and

$$\phi_0 + \frac{\phi_0'}{\lambda} - \frac{a(\theta_0 + \theta_0'/\chi)}{\beta(1 + K_1)} = 0. \qquad (A2.14)$$

For $a \ll \beta$ this gives

$$\phi_0 = -\frac{\phi_0'}{\lambda}. \qquad (A2.15)$$

But since $\lambda \gg \chi$, $(\beta \gg a)$,

$$|\phi_0| = \left|\frac{\phi_0'}{\lambda}\right| \ll \frac{\phi_0'}{\chi}$$

and eqn (A2.13) becomes

$$\theta_0 + \frac{\theta_0'}{\chi} - \frac{\phi_0'}{\chi}\frac{1}{1 + K_{-2}} = 0, \qquad (A2.16)$$

where

$$\chi \simeq Z_D \sqrt{\frac{k_1 + k_{-1}}{D} \cdot \frac{(K_1 + K_{-2} + K_1 K_{-2})}{(1 + K_1)(1 + K_{-2})}}$$

$$= \frac{Z_D}{D^{1/2}} \sqrt{\left(k_1 + \frac{k_{-1}}{1 + K_2}\right)}.$$

In eqn (A2.3)

$$S_1 = 1$$
$$S_2 = 1 + K_2$$
$$\frac{Z_D}{\mu} = \frac{Z_D}{D^{1/2}} \sqrt{\left(k_1 + \frac{k_{-1}}{1 + K_2}\right)} = \chi$$

and substitution for θ_0, θ_0', and ϕ_0' gives eqn (A2.16).

APPENDIX 3

Heterogeneous vs. homogeneous proton transfer

THE second-order rate constant in $cm^3 \, mol^{-1} \, s^{-1}$ for a diffusion controlled reaction is [1]

$$k_{Hom} = 4 \pi L (D_A + D_B)(r_A + r_B),$$

where L is Avogadro's number, and r_A and r_B are molecular radii.

For a heterogeneous process we first work out the target area on the electrode. The concentration of B on the electrode in molecules cm^{-2} is given by

$$[B]_{El} = b K_{ads},$$

where b is the concentration of B in the solution close to the electrode in molecules cm^{-3}, and K_{ads} is the adsorption equilibrium constant in cm. Thus the target area is

$$\pi (r_A + r_B)^2 b K_{ads}.$$

We assume that in the solution there are $1/\lambda^3$ sites for solute molecules per cm^3 and each site is about λcm from another. Then for a concentration of a molecules cm^{-3} the chance of a site being occupied is $a\lambda^3$.

The density of sites per cm^2 is $1/\lambda^2$, and on the Eyring model of diffusion

$$D = \lambda^2 k,$$

where k describes the jumps from one site to the next. Thus the diffusion controlled rate (molecules $cm^{-2} \, s^{-1}$) is

$$\frac{\pi (r_A + r_B)^2 b K_{ads}}{\lambda^2} \times a \lambda^3 \times \frac{k}{2};$$

the factor of $\frac{1}{2}$ allows for jumps in the wrong direction. The heterogeneous rate constant in $cm^4 \, mol^{-1} \, s^{-1}$ is then

$$k_{Het} = \frac{\pi (r_A + r_B) \, K_{ads} \, L D_A}{2 \lambda}.$$

Thus for the two processes we need to compare

$$\frac{D}{\mu} = \sqrt{(D\,k_{\text{Hom}}\,c)}$$

with $k_{\text{Het}}\,c$ where c is the concentration of A in $\text{mol}\,\text{cm}^{-3}$. We can express this by the value of c at which the two processes become equal:

$$c = \frac{8\,\pi\,L\,D^2\,(r_A + r_B) \times 4\,\lambda^2}{\pi^2\,(r_A + r_B)^4\,K_{\text{ads}}^2\,L^2\,D^2}$$

$$\simeq \frac{\lambda^2}{(\bar{r})^3\,L\,K_{\text{ads}}^2}\,,$$

where $r_A + r_B = 2\,\bar{r}$.

For the case where the free energy of a molecule on a site in the solution is no different to a molecule on a site on the electrode we may write

$$K_{\text{ads}} \simeq \lambda,$$

since the number of sites per cm^3 is $1/\lambda^3$, while the number of sites per cm^2 is $1/\lambda^2$.

In general therefore

$$K_{\text{ads}} = \lambda \exp(-\Delta G_{\text{Ads}}/RT),$$

and

$$c = \frac{\exp(2\,\Delta G_{\text{Ads}}/RT)}{\bar{r}^3\,L}\,,$$

where c is the concentration in $\text{mol}\,\text{cm}^{-3}$.

Reference

1. CALDIN, E.F. (1964) In *Fast reactions in solution*, Chap. 1, p. 10. Blackwells, Oxford.

APPENDIX 4

THIS appendix contains expressions for the T_2 term in eqn (9.8).

$$\xi_1'' = \ln(r_2/r_1); \quad \xi_2'' = \ln(r_3/r_2).$$

For $\xi_1'' < \xi_2''$,

$$T_2 = \frac{(\xi_2'')^{5/3}}{\Gamma(8/3)} + \frac{\xi_1''(\xi_2'')^{2/3}}{\Gamma(5/3)} - \frac{(\xi_1'' + \xi_2'')^{5/3}}{\Gamma(8/3)} + \frac{3(\xi_1'' \, \xi_2'')^{1/3}}{10\,[\Gamma(4/3)]^2} \times$$

$$\times \left\{ \xi_1'' - \xi_2'' + (\xi_1'' + \xi_2'')\,\mathcal{F}\left(1,2/3;4/3; \frac{\xi_1''}{\xi_1'' + \xi_2''}\right) \right\}.$$

For $\xi_2'' < \xi_1''$,

$$T_2 = \frac{(\xi_2'')^{5/3}}{\Gamma(8/3)} + \frac{\xi_1'' \, (\xi_2'')^{2/3}}{\Gamma(5/3)} - \frac{3(\xi_1'' \, \xi_2'')^{1/3}}{10\,[\Gamma(4/3)]^2} \times$$

$$\times \left\{ \xi_2'' - \xi_1'' + (\xi_1'' + \xi_2'') \, \mathcal{F}\left(1,2/3;4/3; \frac{\xi_2''}{\xi_1'' + \xi_2''}\right) \right\}.$$

For $\xi_1'' = \xi_2''$, $T_2 = 0{\cdot}718\,\xi''$.

$\mathcal{F}(\ )$ is the hypergeometric function.

APPENDIX 5

Values of a, b, and c in eqn (9.10)

TABLE 1

Values of a

r_3/r_2	r_2/r_1	1·01	1·02	1·03	1·04	1·06	1·08	1·10	1·15	1·20
	1·010	1·06	1·12	1·16	1·19	1·23	1·26	1·27	1·30	1·32
	1·012	1·13	1·18	1·22	1·25	1·30	1·33	1·34	1·36	1·36
	1·014	1·19	1·24	1·28	1·31	1·35	1·38	1·40	1·42	1·42
	1·016	1·25	1·29	1·33	1·36	1·40	1·43	1·45	1·48	1·48
	1·018	1·30	1·34	1·38	1·40	1·45	1·48	1·50	1·53	1·54
	1·020	1·35	1·39	1·42	1·45	1·49	1·52	1·54	1·57	1·59
	1·025	1·46	1·49	1·52	1·55	1·59	1·62	1·64	1·68	1·69
	1·030	1·56	1·58	1·61	1·64	1·68	1·71	1·73	1·77	1·78
	1·035	1·65	1·67	1·69	1·72	1·76	1·79	1·81	1·85	1·87
	1·040	1·73	1·75	1·77	1·79	1·83	1·86	1·89	1·92	1·95
	1·045	1·80	1·82	1·84	1·86	1·90	1·93	1·95	2·00	2·02
	1·050	1·88	1·89	1·91	1·93	1·97	2·00	2·02	2·06	2·09
	1·060	2·01	2·02	2·04	2·05	2·09	2·12	2·14	2·18	2·21
	1·070	2·13	2·13	2·15	2·17	2·20	2·23	2·24	2·27	2·33
	1·080	2·24	2·24	2·26	2·27	2·29	2·31	2·33	2·37	2·39
	1·090	2·34	2·34	2·34	2·35	2·37	2·40	2·42	2·46	2·49
	1·100	2·42	2·41	2·42	2·43	2·45	2·48	2·50	2·54	2·58

APPENDIX 5

TABLE 2

Values of b

r_3/r_2 \ r_2/r_1 →	1·01	1·02	1·03	1·04	1·06	1·08	1·10	1·15	1·20
1·010	0·65	0·75	0·80	0·82	0·83	0·79	0·71	0·50	0·46
1·012	0·74	0·82	0·87	0·89	0·90	0·88	0·83	0·52	0·48
1·014	0·94	1·00	1·04	1·06	1·08	1·06	1·03	0·91	0·68
1·016	1·04	1·09	1·13	1·15	1·17	1·16	1·14	1·03	0·89
1·018	1·12	1·17	1·20	1·23	1·25	1·24	1·22	1·13	1·01
1·020	1·20	1·24	1·27	1·30	1·32	1·31	1·30	1·22	1·11
1·025	1·36	1·39	1·42	1·44	1·47	1·47	1·45	1·39	1·30
1·030	1·50	1·52	1·55	1·57	1·59	1·59	1·58	1·53	1·45
1·035	1·61	1·63	1·66	1·68	1·70	1·70	1·69	1·65	1·58
1·040	1·72	1·73	1·75	1·77	1·79	1·80	1·79	1·75	1·68
1·045	1·81	1·82	1·84	1·85	1·87	1·88	1·88	1·84	1·78
1·050	1·89	1·90	1·92	1·93	1·95	1·96	1·96	1·92	1·87
1·060	2·04	2·04	2·06	2·07	2·09	2·09	2·09	2·04	1·89
1·070	2·18	2·17	2·17	2·19	2·20	2·21	2·19	2·12	2·13
1·080	2·30	2·28	2·28	2·29	2·28	2·26	2·24	2·20	2·13
1·090	2·38	2·35	2·33	2·33	2·32	2·33	2·30	2·27	2·20
1·100	2·43	2·39	2·39	2·39	2·38	2·39	2·36	2·33	2·27

TABLE 3

Values of c

r_3/r_2 ↓ \ r_2/r_1 →	1·01	1·02	1·03	1·04	1·06	1·08	1·10	1·15	1·20
1·010	3·61	2·41	2·04	1·93	2·05	2·58	4·08	115·00	147·00
1·011	2·36	1·77	1·54	1·46	1·51	1·78	2·32	85·00	123·00
1·012	1·76	1·39	1·23	1·16	1·19	1·35	1·66	4·69	113·00
1·013	1·40	1·13	1·00	0·95	0·96	1·07	1·28	2·53	90·20
1·014	1·14	0·94	0·83	0·79	0·79	0·87	1·02	1·79	5·86
1·015	0·96	0·79	0·70	0·66	0·65	0·72	0·83	1·38	2·86
1·016	0·81	0·67	0·59	0·55	0·55	0·60	0·69	1·11	1·99
1·017	0·68	0·57	0·50	0·47	0·46	0·50	0·57	0·91	1·53
1·018	0·58	0·49	0·42	0·39	0·38	0·41	0·48	0·75	1·23
1·019	0·50	0·41	0·36	0·33	0·31	0·34	0·40	0·63	1·01
1·020	0·43	0·35	0·30	0·27	0·26	0·28	0·33	0·53	0·85
1·025	0·17	0·12	0·09	0·07	0·05	0·06	0·09	0·20	0·37
1·030	0·01	−0·02	−0·05	−0·06	−0·08	−0·07	−0·06	0·02	0·13
1·035	−0·10	−0·12	−0·14	−0·16	−0·17	−0·17	−0·16	−0·10	−0·03
1·040	−0·18	−0·20	−0·21	−0·22	−0·24	−0·24	−0·23	−0·19	−0·13
1·045	−0·25	−0·25	−0·27	−0·28	−0·29	−0·29	−0·29	−0·25	−0·21
1·050	−0·30	−0·30	−0·31	−0·32	−0·33	−0·34	−0·33	−0·31	−0·27
1·060	−0·37	−0·37	−0·38	−0·39	−0·40	−0·40	−0·40	−0·37	−0·33
1·070	−0·43	−0·43	−0·43	−0·44	−0·45	−0·45	−0·43	−0·38	−0·35
1·080	−0·48	−0·47	−0·47	−0·48	−0·47	−0·46	−0·44	−0·41	−0·37
1·090	−0·50	−0·49	−0·47	−0·48	−0·47	−0·47	−0·45	−0·43	−0·39
1·100	−0·50	−0·48	−0·48	−0·49	−0·48	−0·48	−0·46	−0·44	−0·41

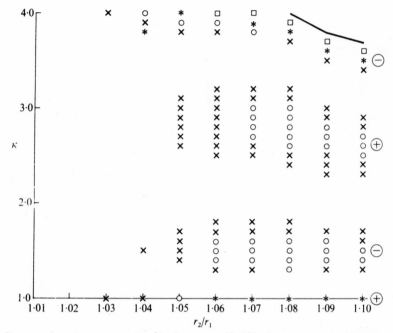

Percentage deviations for N_K from eqn (9.10), for an electrode with $r_3/r_2 = 1{\cdot}06$. Dev $= 100\,(1 - N_{(9.10)}/N_K)$ where $N_{(9.10)}$ is calculated from eqn (9.10). x Dev = 2 to 3%, o Dev = 3 to 4%, *Dev = 4 to 5%, □ Dev = 5 to 7%.

INDEX

Bold type indicates a page on which a literature reference is given